Higher

Geography

2001 Exam
Core
Applications

2002 Exam
Core
Applications

2003 Exam
Core
Applications

2004 Exam
Core
Applications

2005 SQP
Physical and Human Environments
Environmental Interactions

2005 Exam
Physical and Human Environments
Environmental Interactions

Leckie×Leckie

First exam published in 2001.

Published by Leckie & Leckie, 8 Whitehill Terrace, St. Andrews, Scotland KY16 8RN tel: 01334 475656 fax: 01334 477392 enquiries@leckieandleckie.co.uk www.leckieandleckie.co.uk

ISBN 1-84372-341-7

A CIP Catalogue record for this book is available from the British Library.

Printed in Scotland by Scotprint.

Leckie & Leckie is a division of Granada Learning Limited, part of ITV plc.

Acknowledgements

Leckie & Leckie is grateful to the copyright holders, as credited at the back of the book, for permission to use their material.
Every effort has been made to trace the copyright holders and to obtain their permission for the use of copyright material.
Leckie & Leckie will gladly receive information enabling them to rectify any error or omission in subsequent editions.

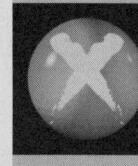

2001 | Higher

[BLANK PAGE]

X042/301

NATIONAL QUALIFICATIONS 2001	WEDNESDAY, 23 MAY 9.00 AM – 10.25 AM	**GEOGRAPHY** HIGHER Core

Attempt **all** questions.

The value attached to each question is shown in the margin.

Credit will be given for appropriate models, diagrams, maps and graphs.

Marks may be deducted for bad spelling, bad punctuation and for writing that is difficult to read.

Note   The reference maps and diagrams in this paper have been printed in black only: no other colours have been used.

SCOTTISH QUALIFICATIONS AUTHORITY

©

Extract No 1216/124

1:50 000 Scale
Landranger Series

1 mile = 1·6093 kilometres

Marks

Question 1

Study Reference Diagram Q1.

Suggest **physical** and **human** factors which might have contributed to the variations in temperature shown in Reference Diagram Q1.

5

Reference Diagram Q1 (Variations in average world temperatures 1850–1990)

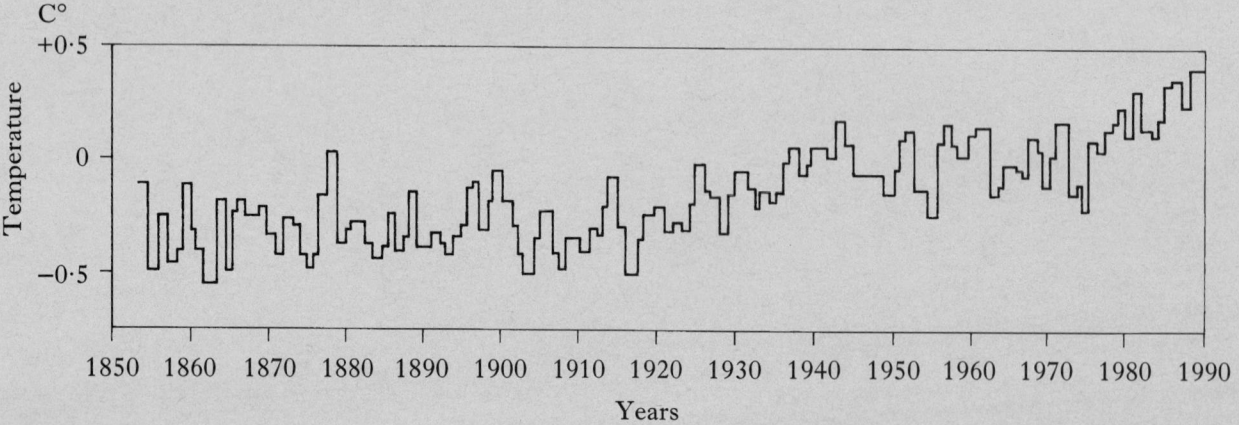

Marks

Question 2

Study OS map extract number 1216/124: Dolgellau (*separate item*).

Using appropriate grid references, **describe** the **physical characteristics** of the Afon (River) Dysynni and its valley from 710094 to 608050 (where it "leaves" the map extract).

4

Question 3

Study OS map extract number 1216/124: Dolgellau (*separate item*), and Reference Map Q3.

(*a*) The area within Area A, shown on Reference Map Q3, has been greatly affected by glacial erosion.

Identify **two** different features of glacial erosion in this part of the map extract, and give their grid references.

2

(*b*) Choose **one** of these features of glacial erosion and, with the aid of annotated diagrams, **explain** how it was formed.

4

Reference Map Q3

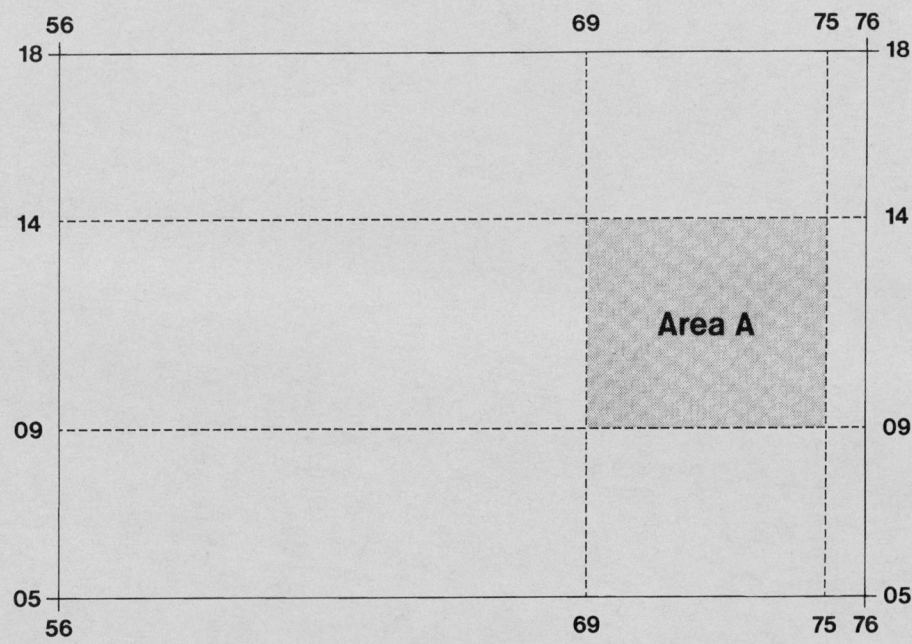

[Turn over

Marks

Question 4

Study Reference Diagram Q4 which shows information collected from a field survey of a coastal sand dune transect.

Suggest reasons for the changes in vegetation along the line of the transect. You should refer to a range of environmental factors.

5

Reference Diagram Q4 (Main plant types along a coastal sand dune transect)

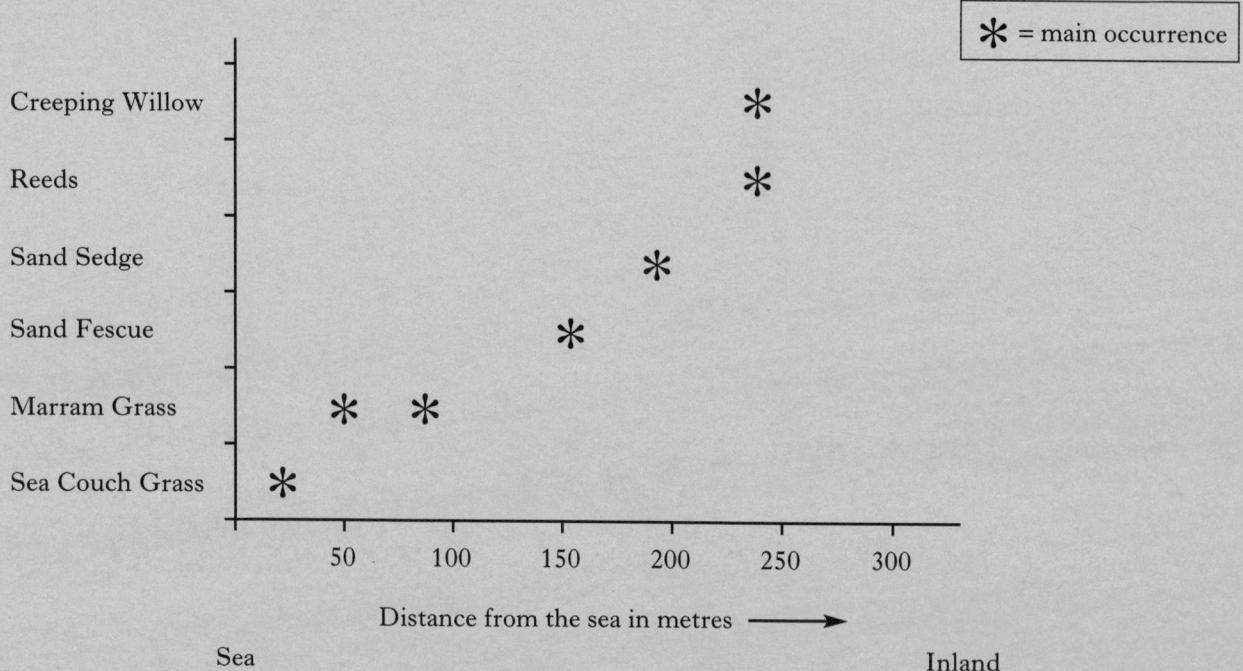

Marks

Question 5

Study Reference Diagram Q5.

India has a population structure which is typical of that of many **Developing** Countries. **Describe** and **account for** the population structure shown.

5

Reference Diagram Q5 (India: population pyramid 1991)

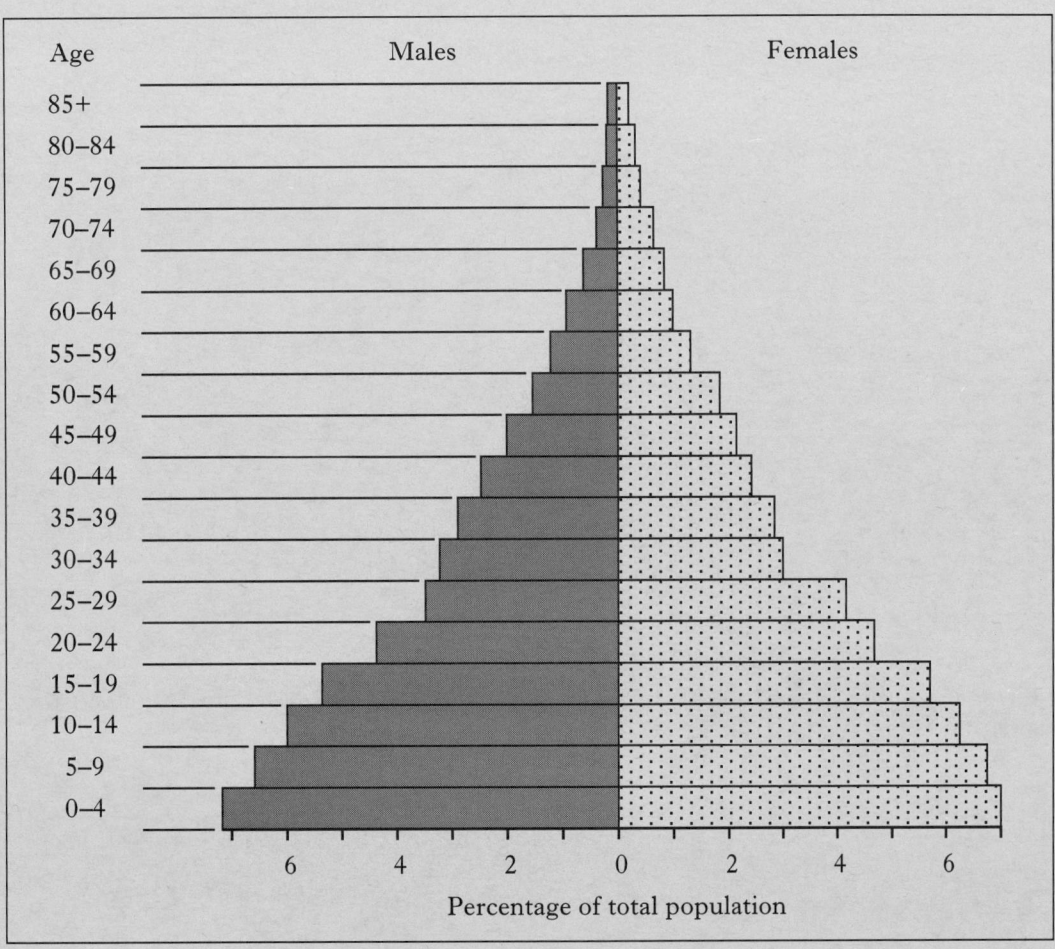

[Turn over

Marks

Question 6

Study Reference Diagram Q6.

For this area, or any area of intensive peasant farming you have studied, **describe** and **account for** the main features of the farming landscape.

5

Reference Diagram Q6 (An intensive peasant farming landscape in Southern Asia)

Marks

Question 7

Study Reference Diagram Q7.

With reference to **one named** industrial concentration in the European Union which you have studied, **explain** how such factors **originally** attracted industry to your chosen area.

5

Reference Diagram Q7 (Factors affecting the location of industry)

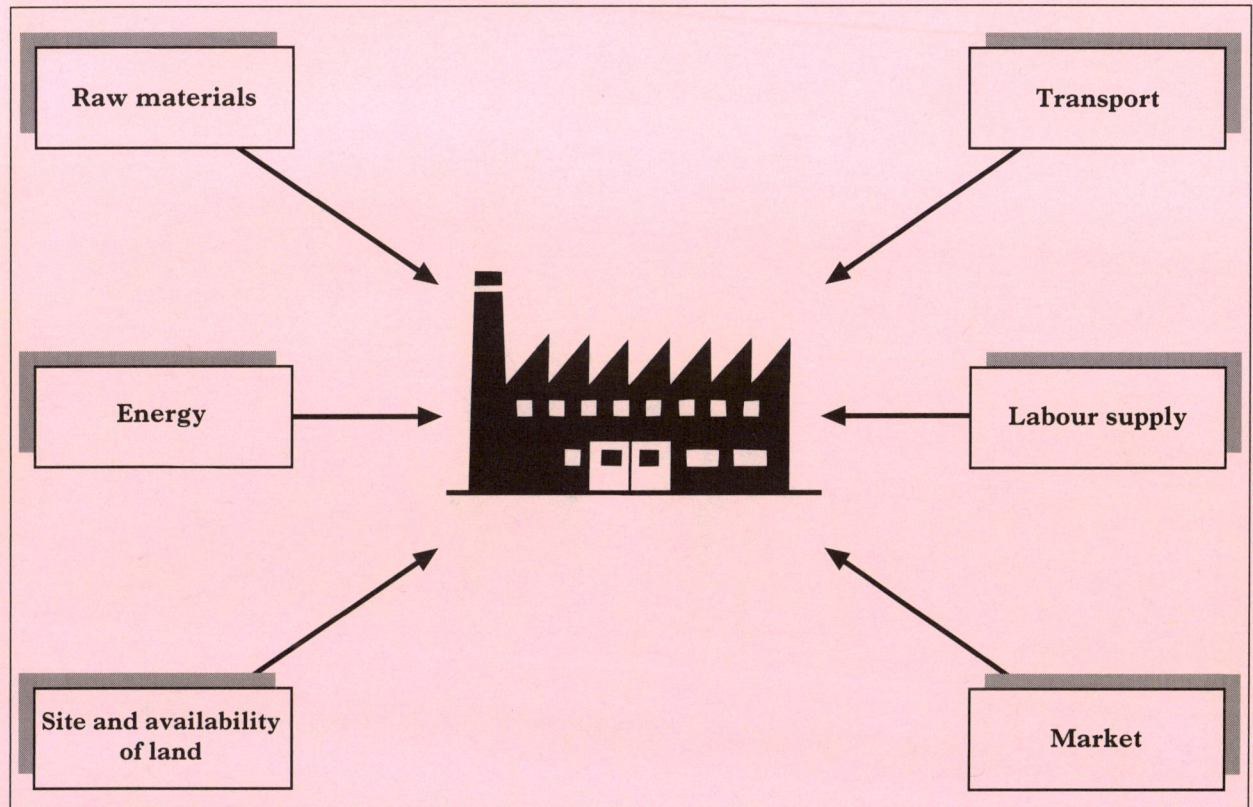

[Turn over for Question 8 on *Page eight*

Marks

Question 8

"*The Central Business District of major cities undergoes continuing change.*"

Referring to a city that you have studied in the **Developed** World, **explain** the changes which have taken place in the CBD over the past few decades.

You should refer to named locations within the CBD.

5

[END OF QUESTION PAPER]

NATIONAL QUALIFICATIONS 2001	WEDNESDAY, 23 MAY 10.45 AM – 12.05 PM	**GEOGRAPHY** HIGHER Applications

Two questions should be attempted.

One question from Section 1 (Questions 1, 2, 3) and **one** question from Section 2 (Questions 4, 5, 6).

Write the numbers of the **two** questions you have attempted in the marks grid on the back cover of your answer booklet.

The value attached to each question is shown in the margin.

Credit will be given for appropriate models, diagrams, maps and graphs.

Marks may be deducted for bad spelling, bad punctuation and for writing that is difficult to read.

Note The reference maps and diagrams in this paper have been printed in black only: no other colours have been used.

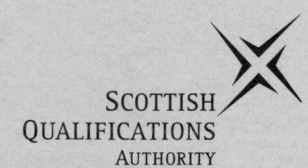

SCOTTISH QUALIFICATIONS AUTHORITY

Marks

SECTION 1

You must answer ONE question from this Section.

Question 1 (Rural Land Resources)

(a) Study Reference Map Q1.

Upland areas of the UK include areas of **Carboniferous Limestone** landscape and Chalk and Jurassic Limestone uplands characterised by their **Scarp and Vale** landscape.

Choose **one** of these types of landscape and, with the aid of annotated diagrams, **explain** how the main features of the physical landscape were formed.   **10**

(b) For any **named** upland area of the UK which you have studied, **explain** the main social and economic opportunities provided by the landscape.   **6**

(c) For your chosen upland area,

    (i) **give examples** of environmental conflicts which have arisen, and

    (ii) **describe** some of the measures taken to resolve these conflicts and **comment** on their effectiveness.   **9**

          (25)

Reference Map Q1 (Chalk and Limestone upland areas in England)

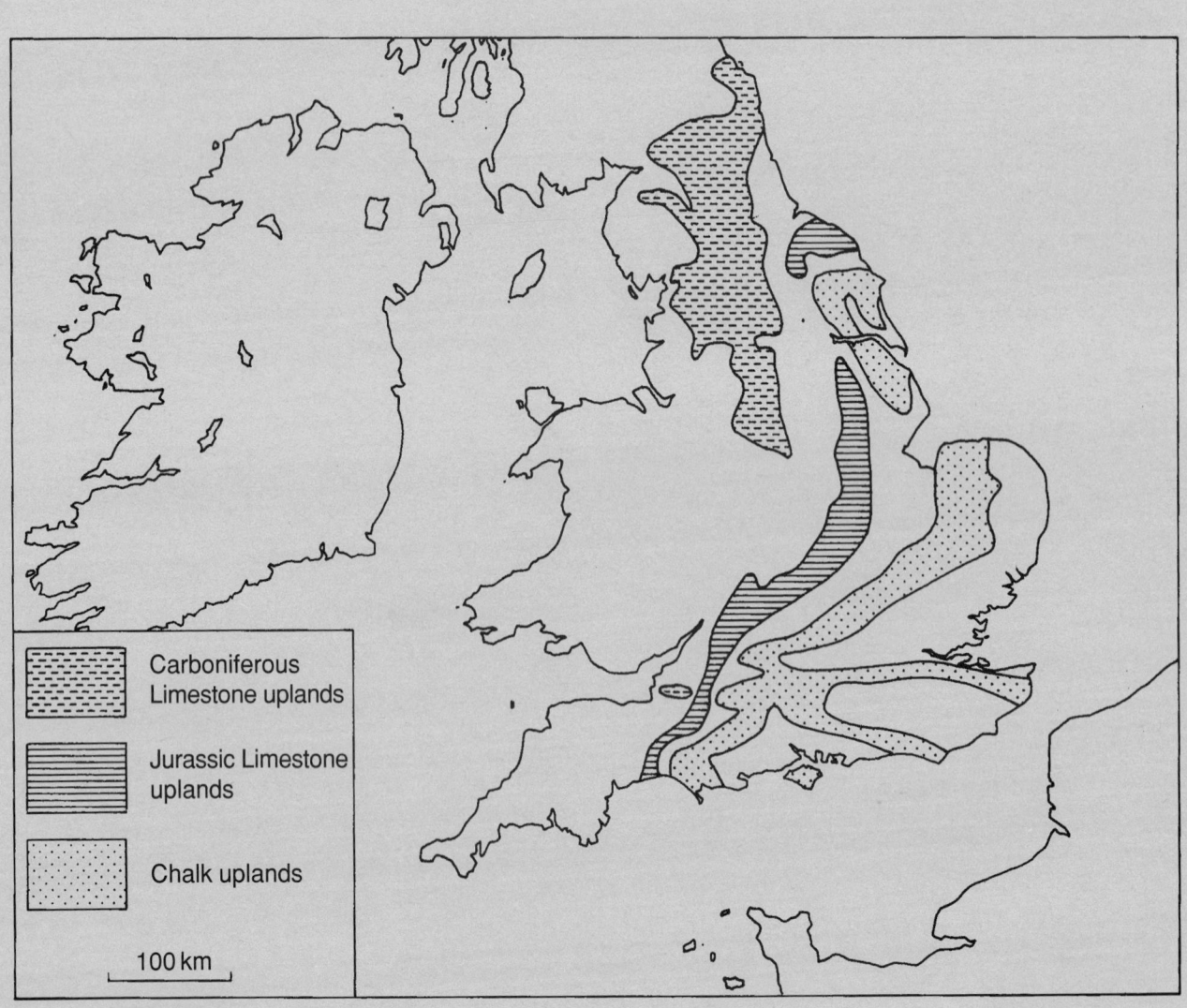

Carboniferous Limestone uplands

Jurassic Limestone uplands

Chalk uplands

100 km

Marks

Question 2 (Rural Land Degradation)

(*a*) Study Reference Diagram Q2.

Describe in detail the **physical factors** which have led to land degradation in named areas of North America **and either** Africa North of the Equator **or** the Amazon Basin.   **7**

(*b*) Referring to named locations in **either** Africa North of the Equator **or** the Amazon Basin, **describe** the social, economic and environmental impact of rural land degradation.   **9**

(*c*) Referring to named areas of North America, **describe** the measures which have been taken to try to conserve soil and reduce land degradation, and **comment** on their effectiveness.   **9**

    (25)

Reference Diagram Q2 (Physical factors which can lead to land degradation)

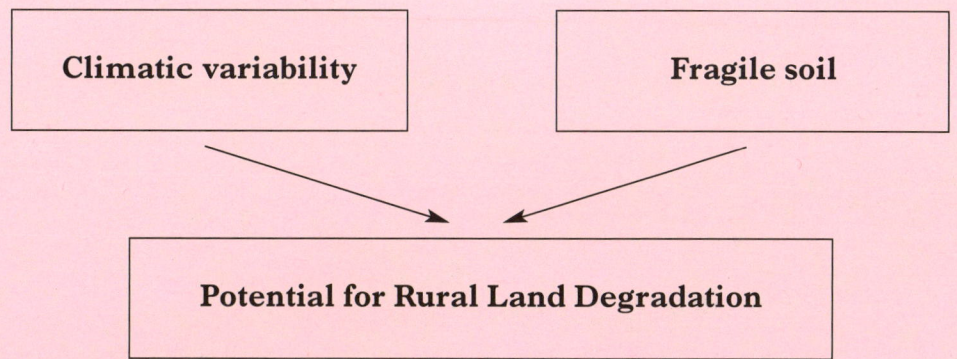

[Turn over

Marks

Question 3 (River Basin Management)

(*a*)   Study Reference Map Q3 and Reference Diagram Q3A which show the location of dams in the Tennessee Valley Authority.

With reference to the Tennessee Valley **or** any other river basin you have studied in North America **or** Africa, **explain** the **human and physical factors** that have to be considered when selecting sites for dams and their associated reservoirs.   **7**

(*b*)   Study Reference Diagram Q3B.

With reference to your chosen river basin, **explain** how water management projects have affected the hydrological cycle of the river basin.   **5**

(*c*)   For your chosen river basin, **explain**

   (i)   why there was a need for a water management project, and

   (ii)   the benefits which were gained from the water management project.   **13**

(25)

Reference Map Q3 (The Tennessee Valley Authority area)

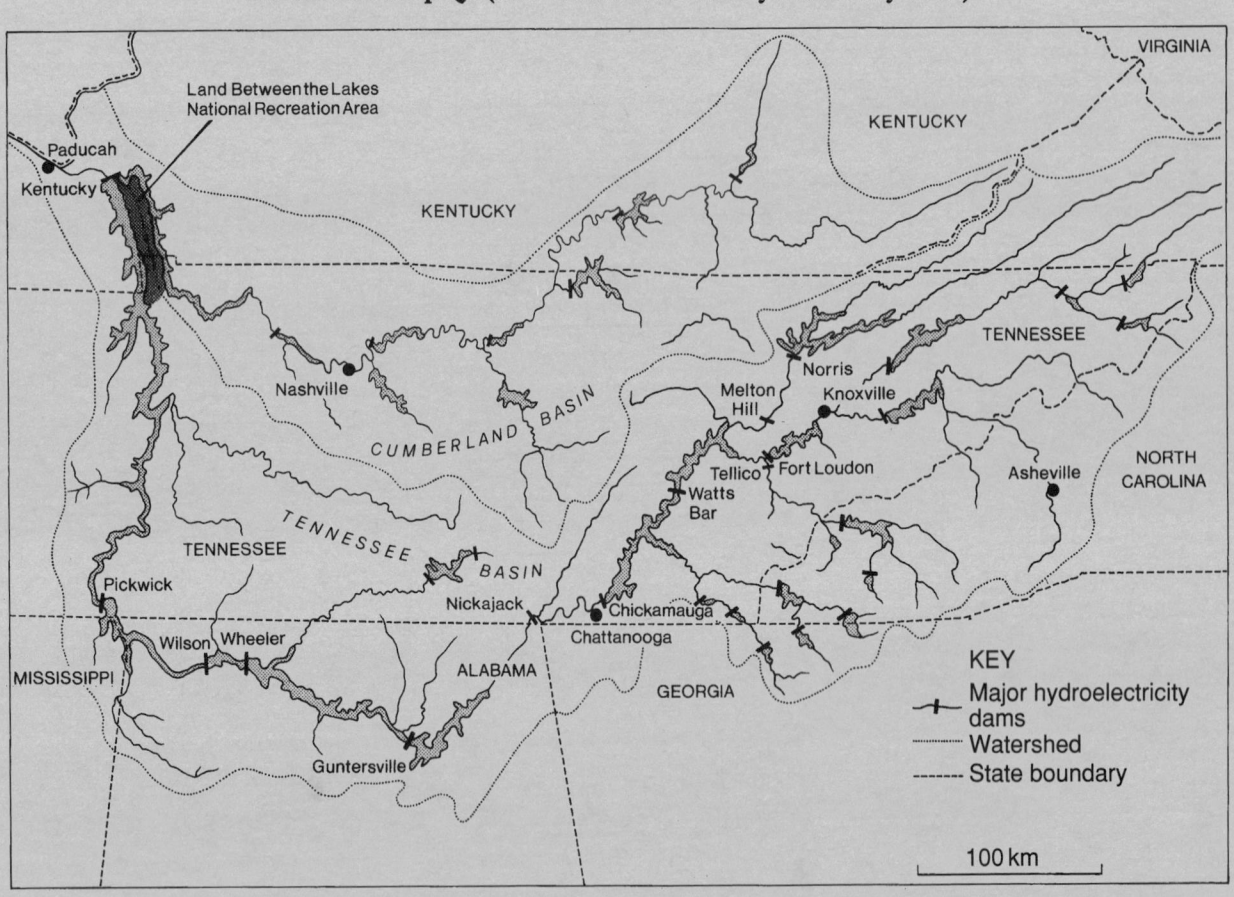

Question 3 – continued

Reference Diagram Q3A (Location of major dams on the Tennessee River)

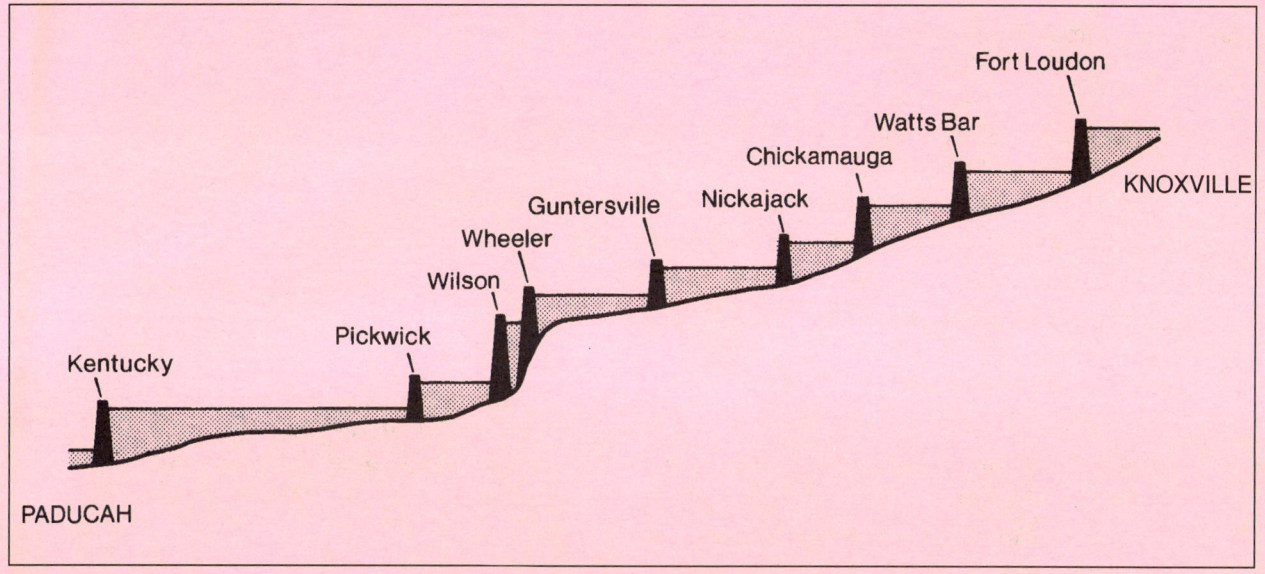

Reference Diagram Q3B (The hydrological cycle)

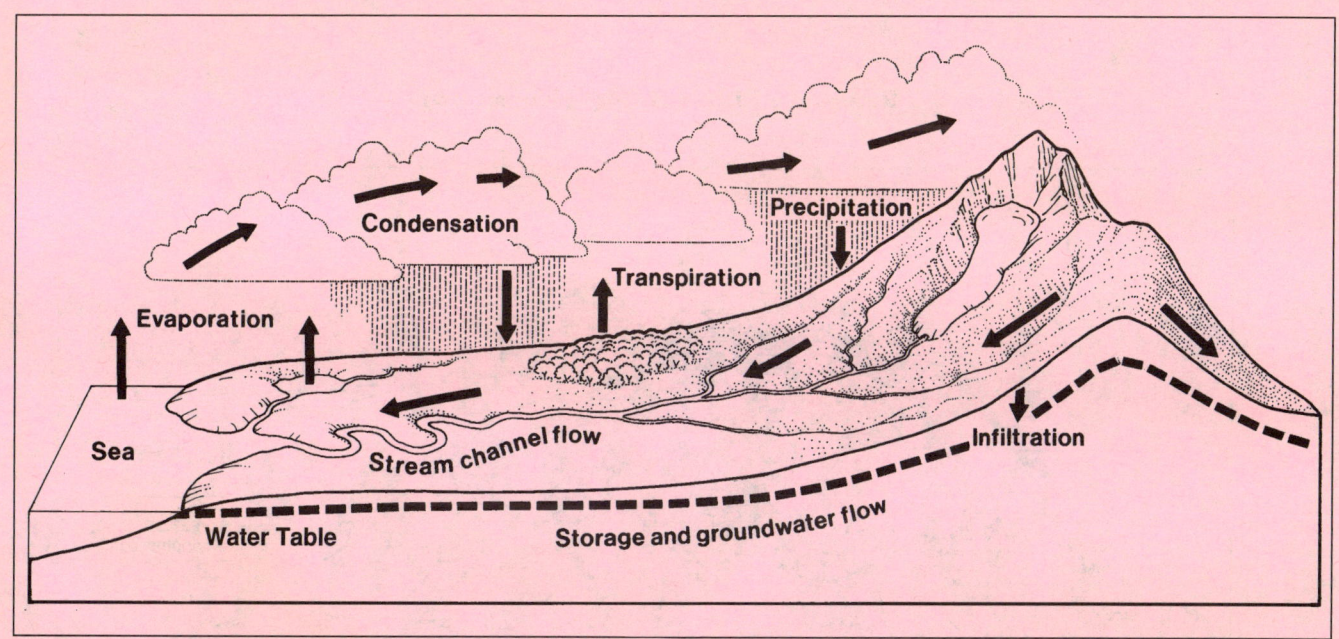

[Turn over

Marks

SECTION 2

You must answer ONE question from this Section.

Question 4 (Urban Change and its Management)

(*a*)   Study Reference Diagram Q4 which highlights the problem of urban sprawl, faced by many countries in the Developed World.

With reference to a **named** city in the **Developed** World,

   (i)   **describe** some of the problems caused by the continued expansion of the city,

   (ii)   **describe** strategies which have been introduced to halt the spread of the city, and

   (iii)   **comment** on the effectiveness of these strategies.   **10**

(*b*)   Study Reference Map Q4.

"*Delhi has shanties scattered across the city rather than restricted to the periphery.*"

For Delhi, **or** a **named** city you have studied in the **Developing** World, **describe** the factors which have

   (i)   led to the growth of shanty towns, and

   (ii)   influenced the location of shanty towns.   **8**

(*c*)   "*Many cities in the Developing World are taking steps to improve the quality of life for their residents.*"

With reference to the city you have chosen in (*b*), **describe** the schemes designed to improve the quality of life in the city, and **comment** on their effectiveness.   **7**

                                                                                                                    (25)

Reference Diagram Q4 (Urban sprawl)

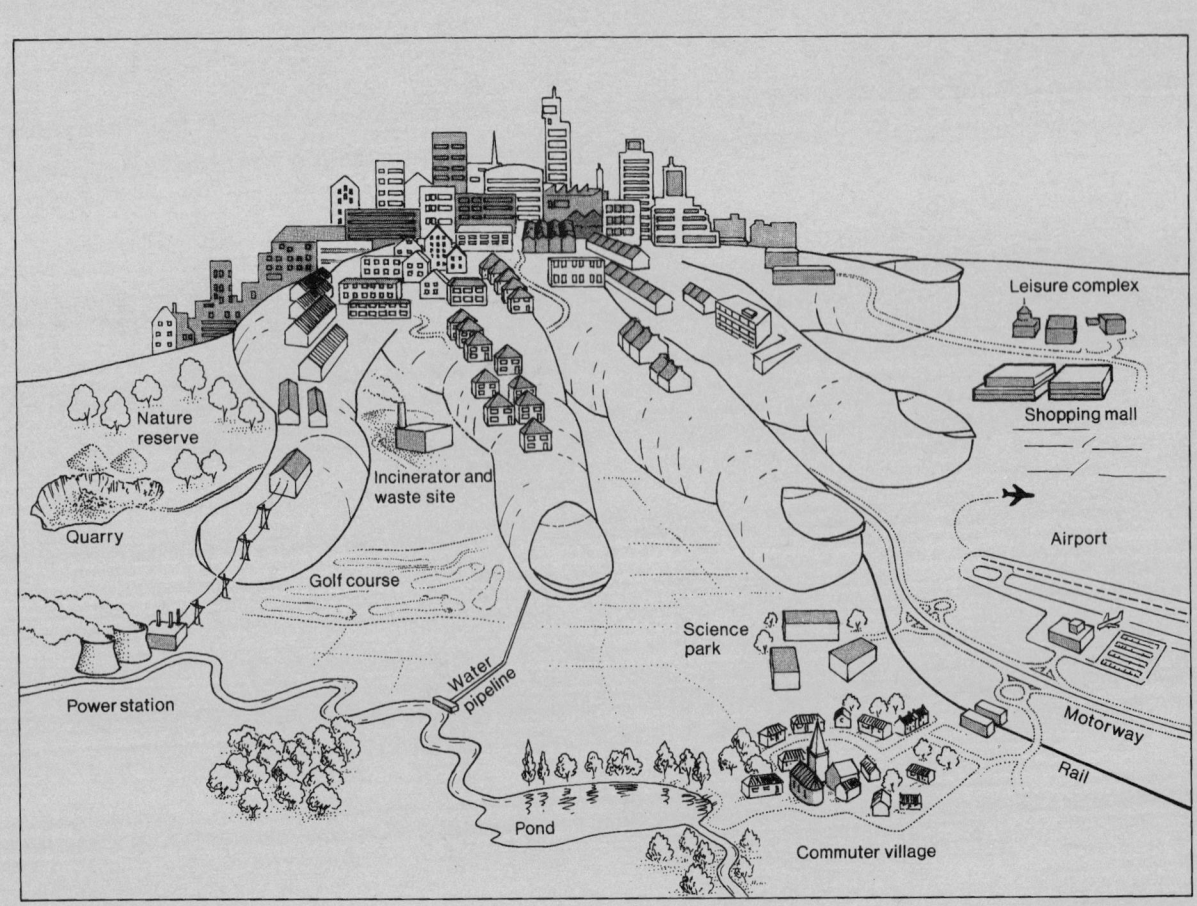

Question 4 – continued

Reference Map Q4 (Distribution of shanties in Delhi)

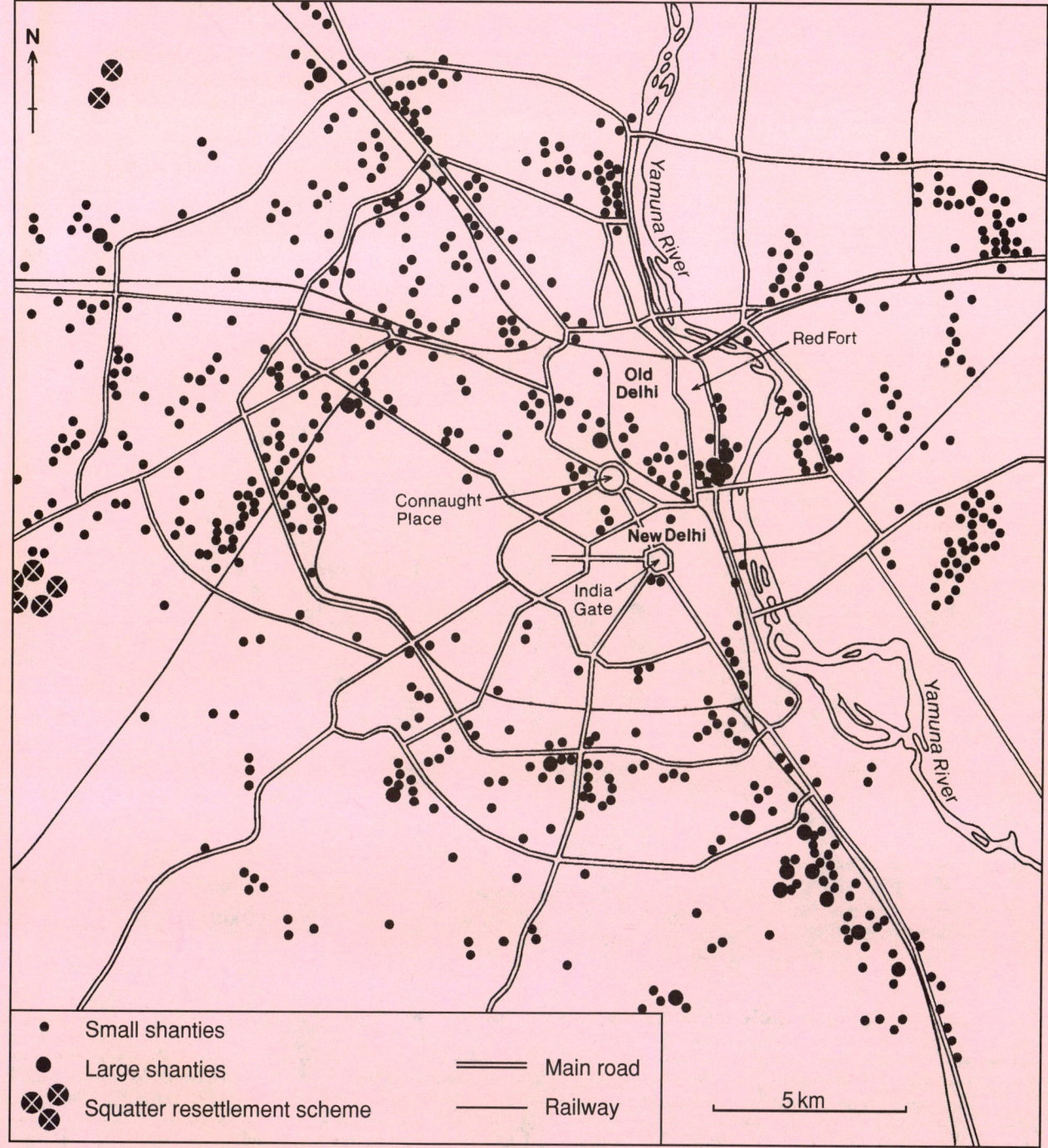

[Turn over

Marks

Question 5 (European Regional Inequalities)

(a) Study Reference Map Q5 and Reference Table Q5.

In what ways does the data provide evidence of regional inequalities within Spain? **5**

(b) For **either** Spain **or** a named country you have studied in the European Union, **describe** and **explain** both the physical **and** human factors which have led to regional inequalities. **10**

(c) For **either** Spain **or** a named country you have studied in the European Union,

   (i) **describe** the steps taken by the national government and the European Union to tackle problems in the less prosperous regions, and

   (ii) **comment** on the effectiveness of these steps. **10**

(**25**)

Reference Map Q5 (Spanish Regions)

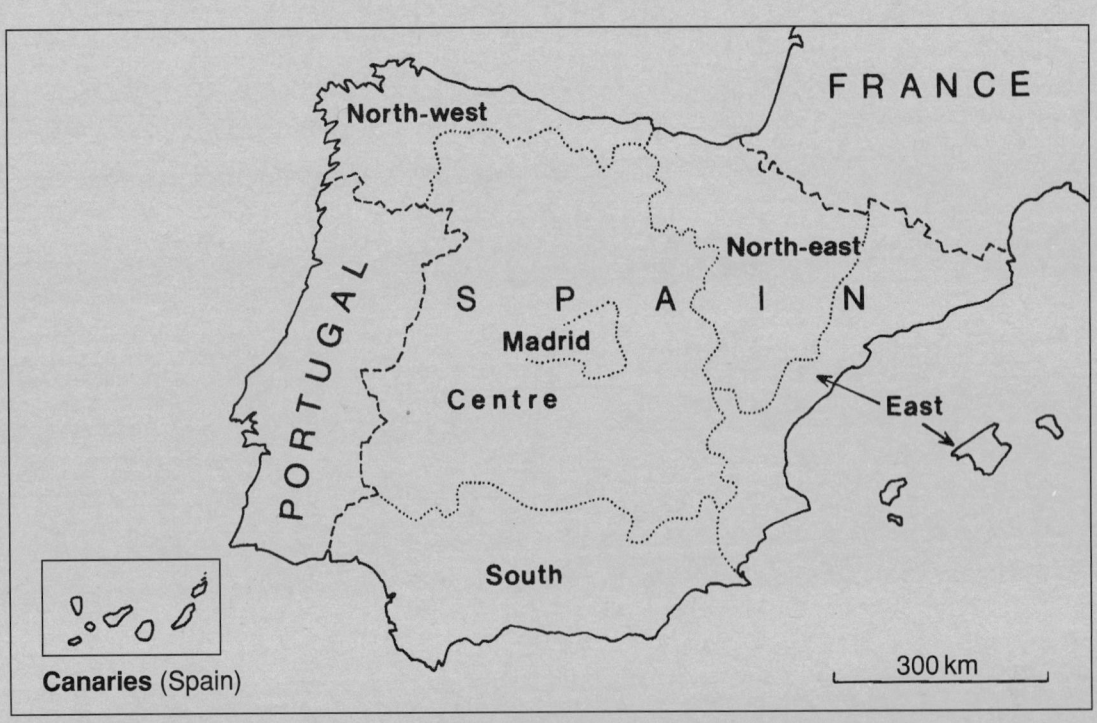

Reference Table Q5 (Socio-economic data for the regions of Spain)

Region	Area (% of total)	Population (% of total)	Percentage of Population		Employment % in			Unemployment Rate (%)	GDP per hea (EU = 100)
			Aged under 15	Aged 65 and over	Agriculture	Industry	Services		
Spain			**16·9**	**15·1**	**8·6**	**29·4**	**62·0**	**22·3**	**77**
North-west	8·9	10·5	14·7	17·8	21·4	27·2	51·4	20·4	65
North-east	14·0	9·7	14·2	16·5	6·4	35·1	58·5	17·9	91
Madrid	1·6	15·0	16·3	13·3	1·0	26·0	73·0	20·6	96
Centre	42·6	13·0	16·5	18·4	14·3	29·4	56·3	22·2	65
East	11·9	26·0	16·2	15·3	4·6	34·9	60·5	19·4	89
South	19·6	22·1	20·3	12·7	10·9	22·9	66·2	31·3	59
Canaries	1·4	3·7	19·6	10·4	8·3	19·2	72·5	21·7	75

Marks

Question 6 (Development and Health)

(*a*)　Study Reference Table Q6.

　　(i)　**Suggest reasons** for the differences in development between Newly Industrialising Countries and other Developing Countries. You should refer in your answer to Newly Industrialising Countries and Developing Countries you have studied. **6**

　　(ii)　**Explain** why indicators of development such as those used in Reference Table Q6 may fail to reflect accurately the true quality of life **throughout** a country. **4**

(*b*)　Study Reference Map Q6 which shows the main areas of the world at risk from cholera.

　　Referring to cholera, **or** malaria, **or** bilharzia/schistosomiasis,

　　(i)　**describe** the physical **and** human factors which put people at risk of contracting the disease,

　　(ii)　**describe** and **evaluate** the strategies used in controlling the spread of the disease, and

　　(iii)　**explain** the benefits to Developing Countries of controlling the disease. **15**

(25)

Reference Table Q6 (Indicators of development for two countries)

Indicators	Sample Newly Industrialising Country South Korea	Sample Developing Country Sudan
People per doctor	1176	11 364
Life expectancy (years)	70	50
GDP per capita (US$)	4081	467

Reference Map Q6 (Countries with recent cholera outbreak)

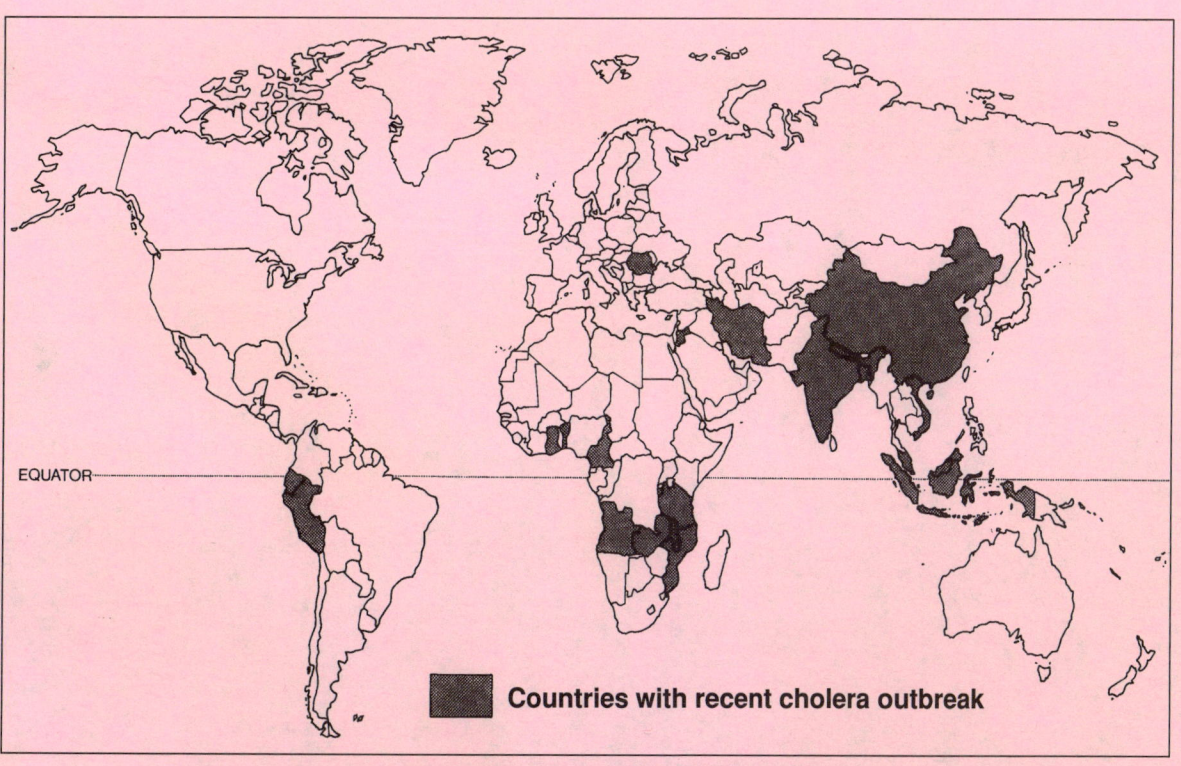

EQUATOR

■ **Countries with recent cholera outbreak**

[END OF QUESTION PAPER]

[BLANK PAGE]

2002 | Higher

[BLANK PAGE]

NATIONAL QUALIFICATIONS 2002	WEDNESDAY, 5 JUNE 9.00 AM – 10.30 AM	GEOGRAPHY HIGHER Core

Attempt **all** questions.

The value attached to each question is shown in the margin.

Credit will be given for appropriate models, diagrams, maps and graphs.

Marks may be deducted for bad spelling, bad punctuation and for writing that is difficult to read.

Note   The reference maps and diagrams in this paper have been printed in black only: no other colours have been used.

SCOTTISH QUALIFICATIONS AUTHORITY

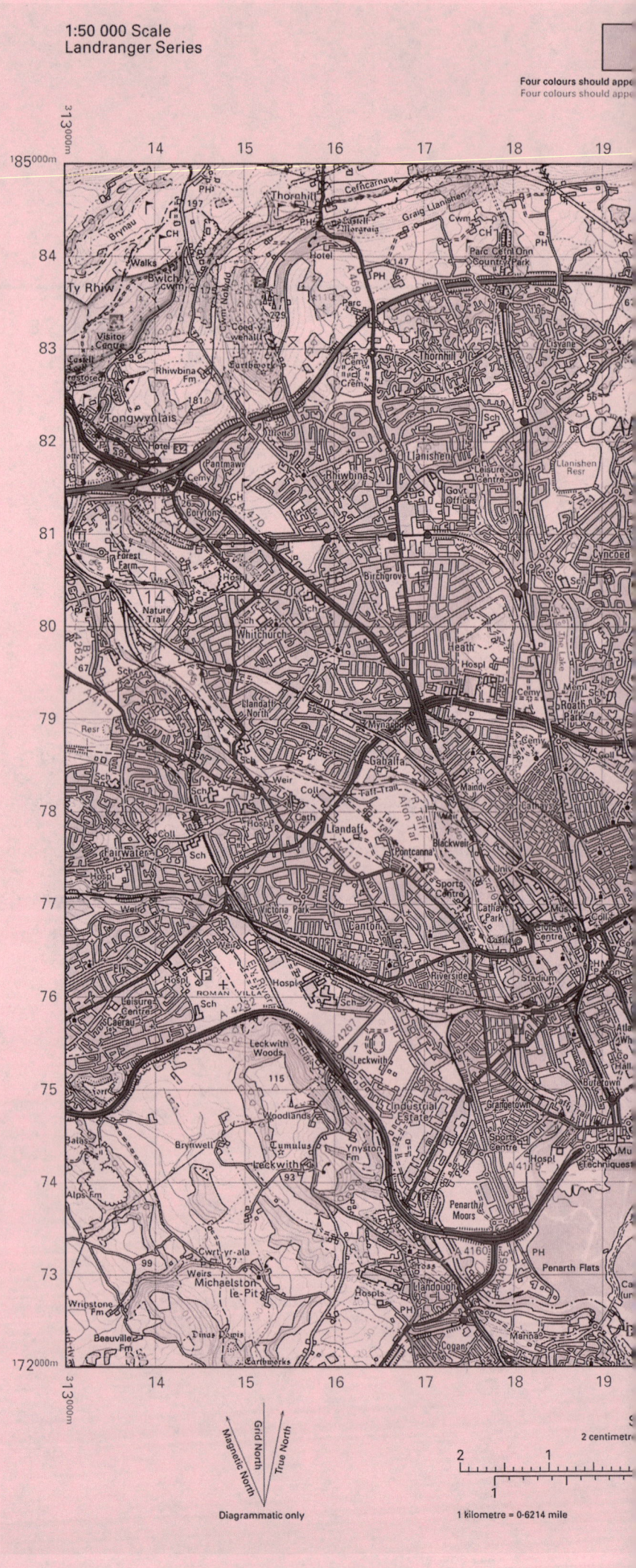

Extract No 1269/171

Marks

Question 1

Study Reference Diagram Q1.

(a)   **Describe** the latitudinal variation of the Earth's energy balance shown in the diagram.   2

(b)   With the aid of an annotated diagram of the Earth, **explain** the variations.   4

Reference Diagram Q1 (Latitude and energy balance)

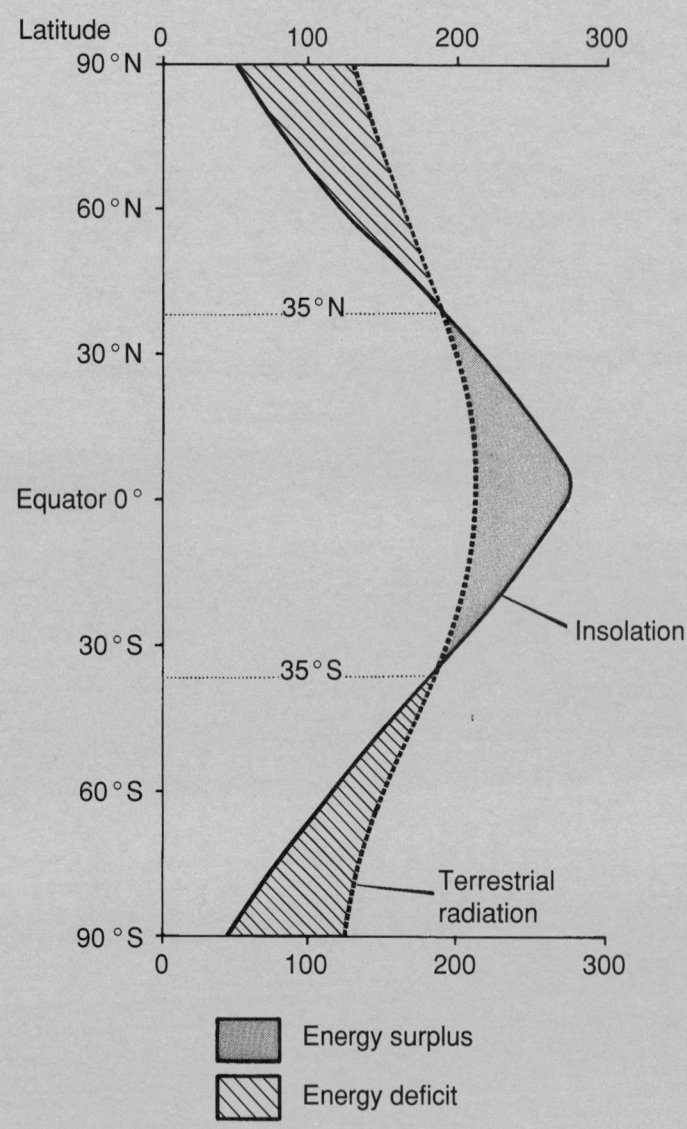

Marks

Question 2

Study Reference Diagram Q2 and Reference Maps Q2.

(*a*) **Describe** the variations throughout the year in the flow of the river Niger at Mopti.   **2**

(*b*) Using the diagram and the maps, **suggest reasons** for the variations in the flow.   **4**

Reference Diagram Q2 (Hydrograph and precipitation—River Niger at Mopti)

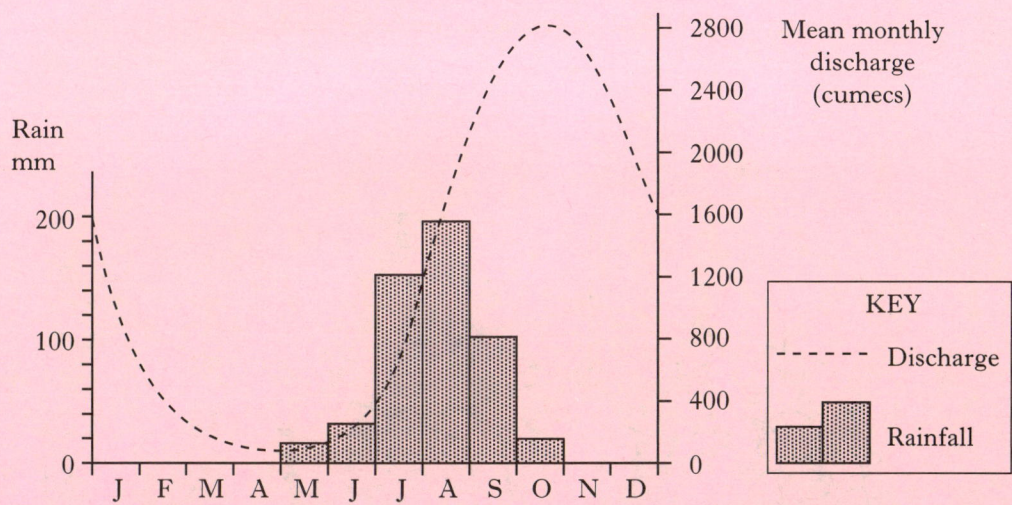

Reference Maps Q2 (Selected air masses and fronts over Africa in January and July)

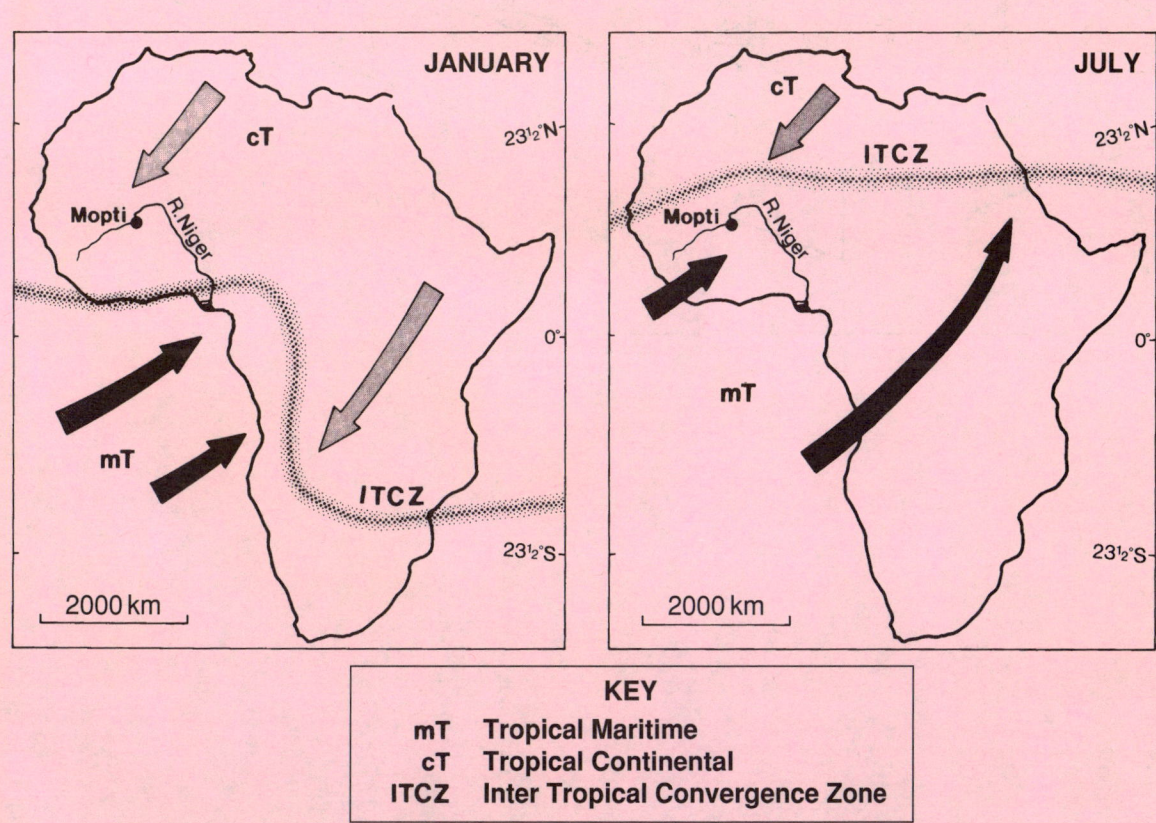

KEY

mT   **Tropical Maritime**
cT   **Tropical Continental**
ITCZ   **Inter Tropical Convergence Zone**

Question 3

Study Reference Diagram Q3.

Select **two** features from the following list and **explain** the processes involved in their formation:

  (i)   limestone pavement;

  (ii) gorge;

  (iii) stalactites and stalagmites.

7

Reference Diagram Q3 (Carboniferous Limestone landscape)

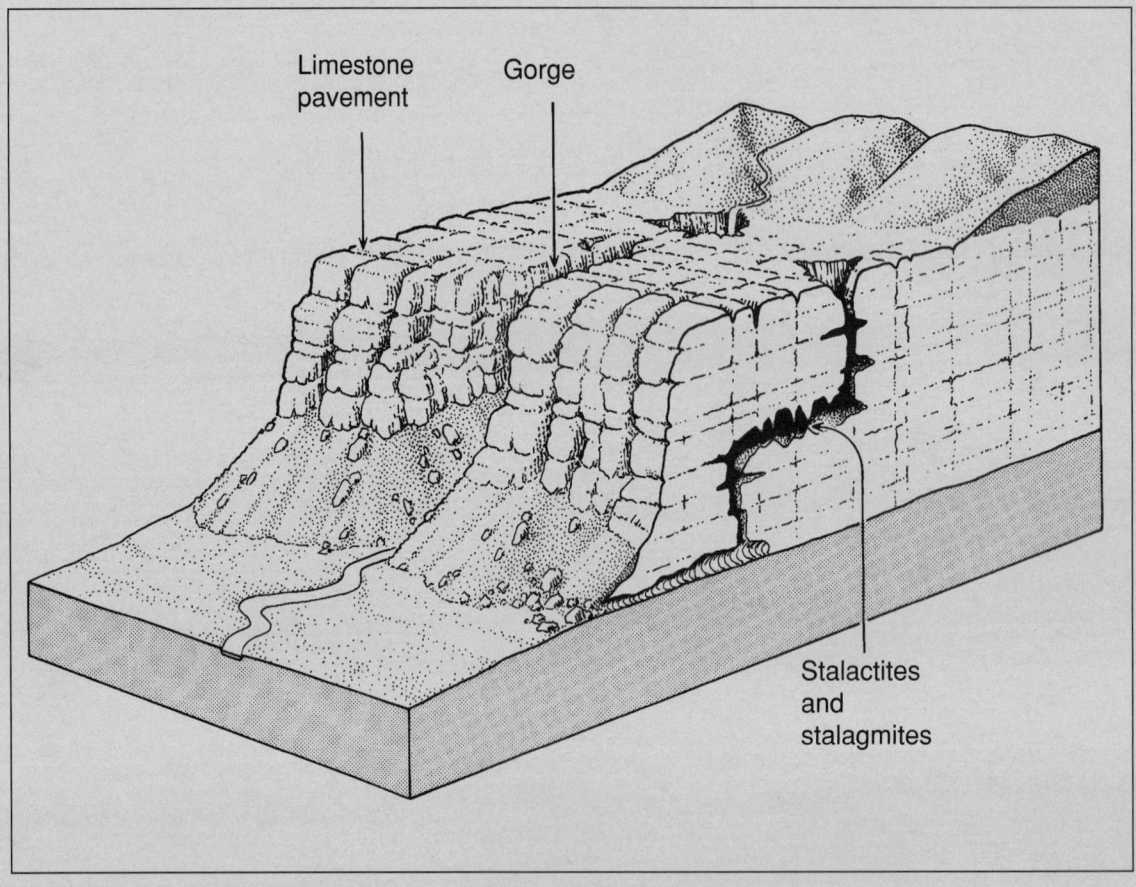

Marks

Question 4

Study Reference Diagram Q4.

Select **one** of the following soil types:

  (i)   gley;

 (ii)   podzol;

(iii)   brown earth.

With the aid of an annotated sketch of a soil profile, **explain how** the major soil forming factors shown in the diagram have contributed to its formation.

6

Reference Diagram Q4 (Main factors affecting soil formation)

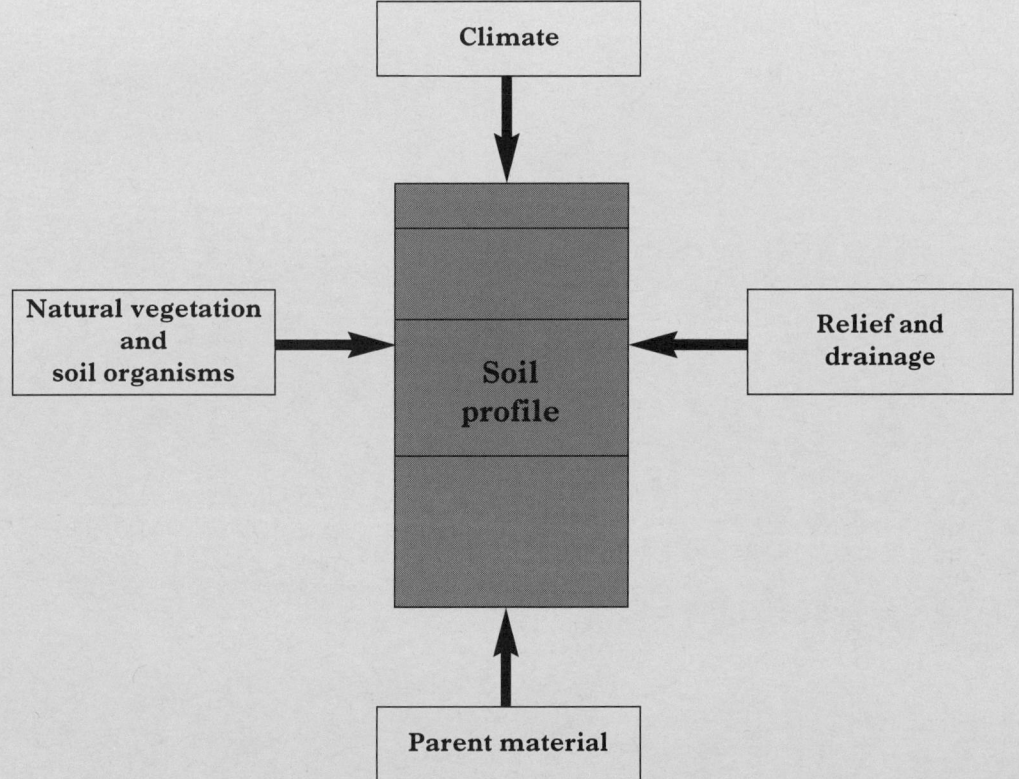

[Turn over

Marks

Question 5

Study Reference Diagram Q5.

International migrations may be **voluntary** or **forced**.

Referring to **one named** example of **each** type of migration, **explain** why the migration took place.

6

Reference Diagram Q5 (A model of migration)

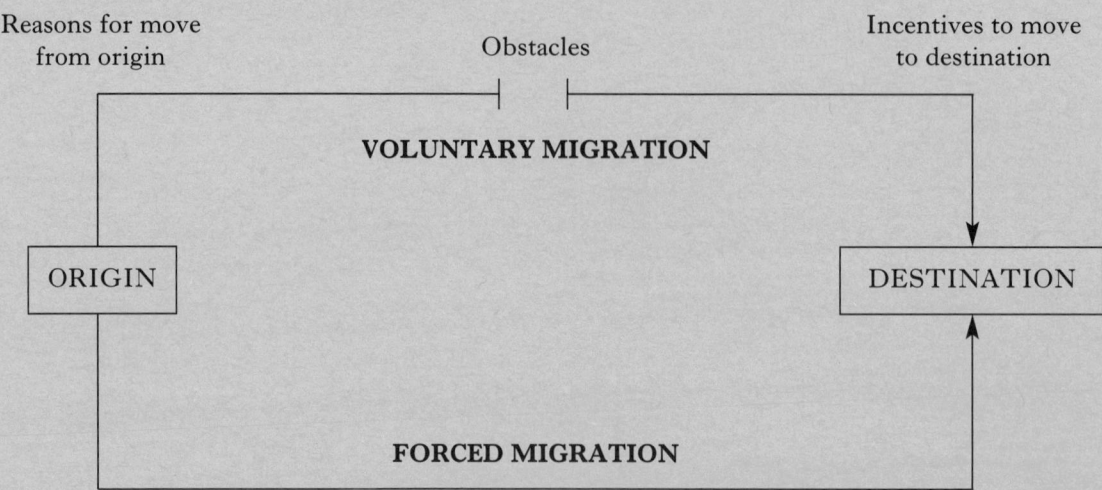

Marks

Question 6

Study Reference Diagram Q6 which shows the relative importance of the elements of shifting cultivation.

"Shifting cultivation remains an important farming system in many Tropical areas."

For a named location, **describe** and **explain** the characteristics of this farming system.

6

Reference Diagram Q6 (Elements of the system of shifting cultivation)

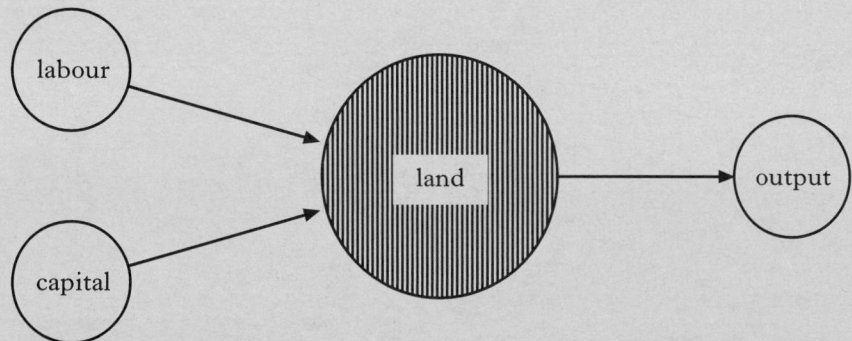

[Turn over

Marks

Question 7

Study OS map extract number 1269/171: Cardiff (*separate item*), and Reference Map Q7.

Using map evidence, **describe** and **explain** the physical and human factors which encouraged industry to locate in Area A.

6

Reference Map Q7

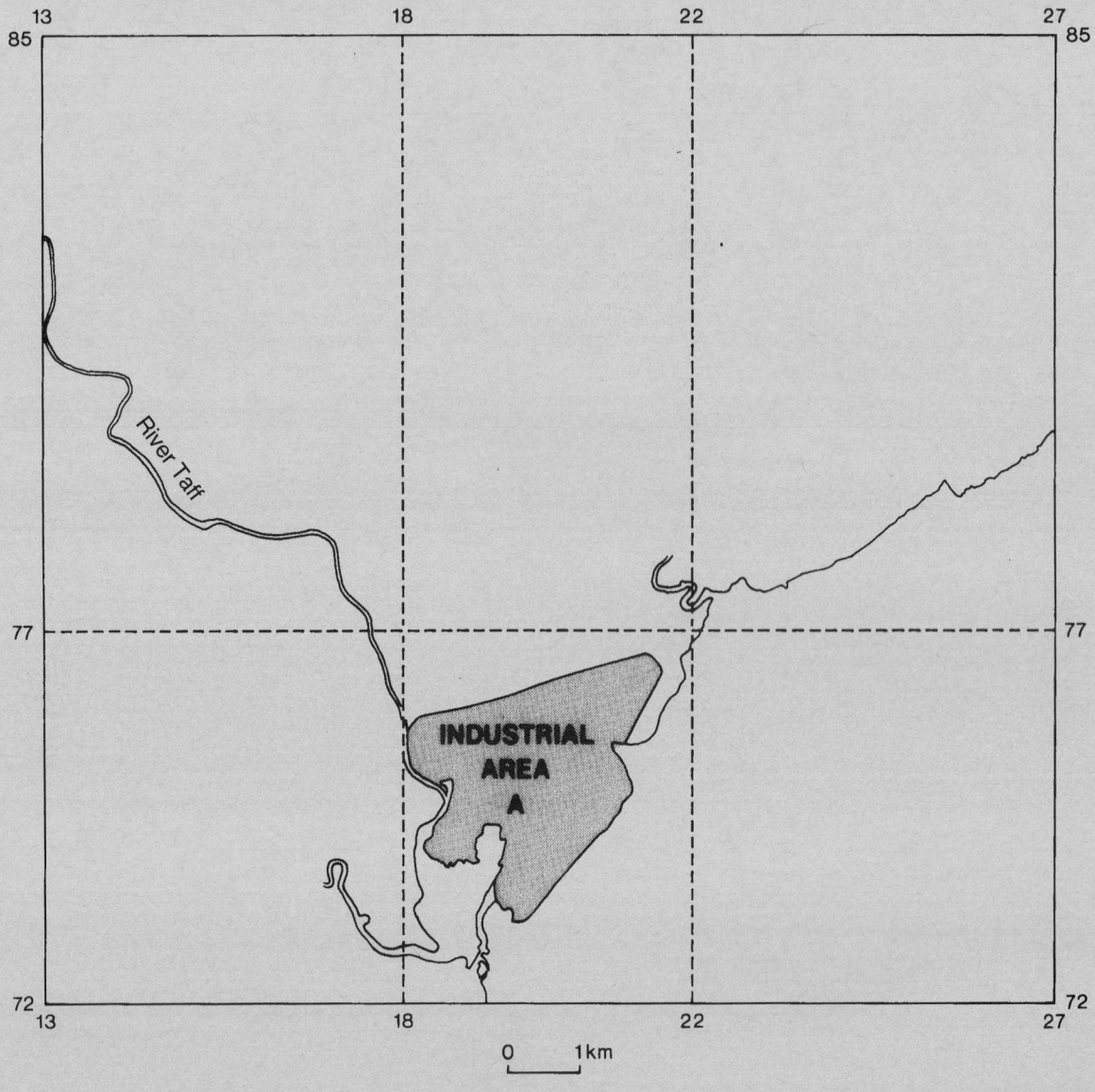

Marks

Question 8

Study OS map extract number 1269/171: Cardiff (*separate item*), and Reference Map Q8.

Describe the urban environments of Areas X and Y and **suggest reasons** for the differences.

7

Reference Map Q8

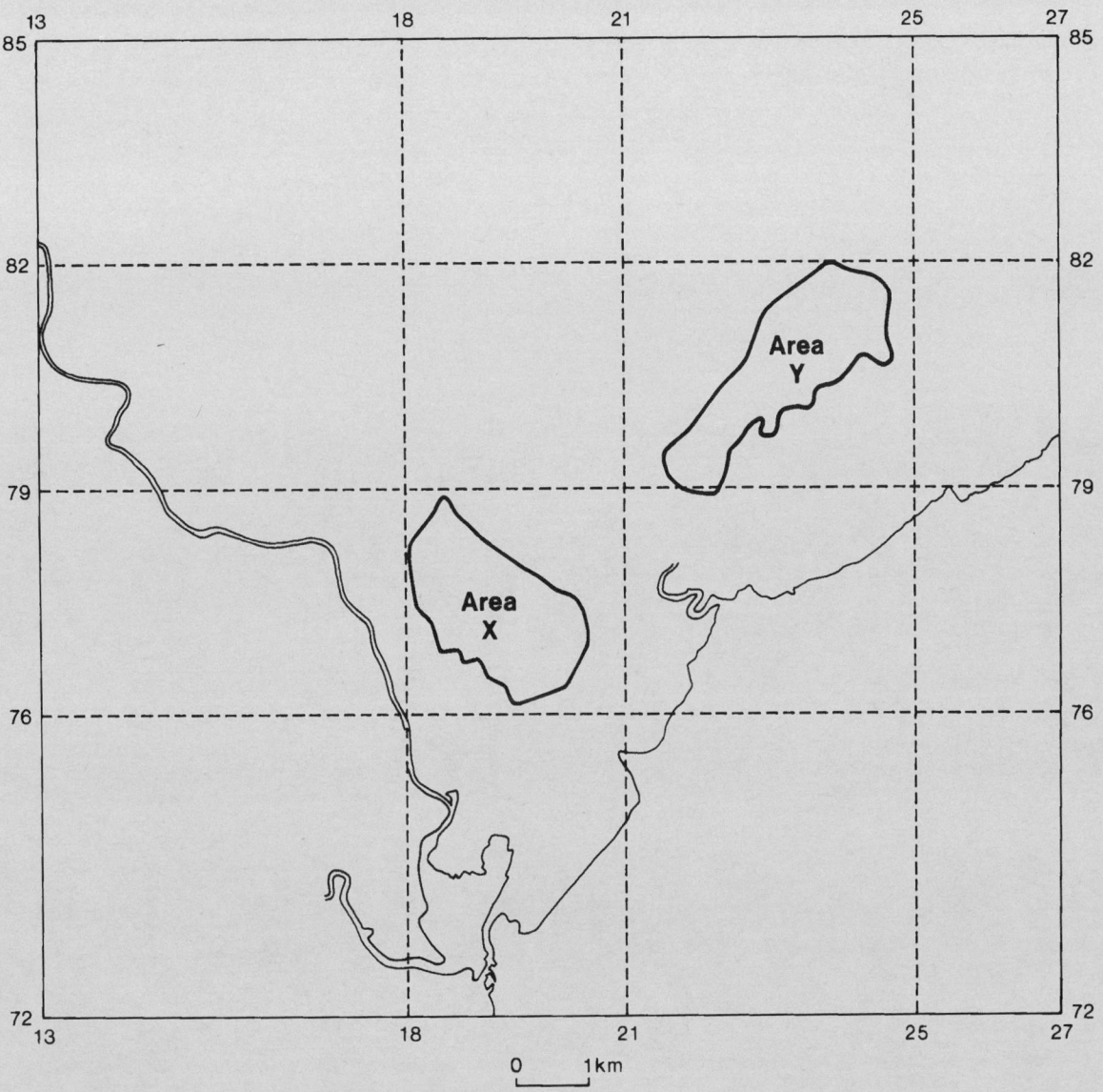

0   1km

[*END OF QUESTION PAPER*]

[BLANK PAGE]

NATIONAL QUALIFICATIONS 2002	WEDNESDAY, 5 JUNE 10.50 AM – 12.05 PM	**GEOGRAPHY** HIGHER Applications

Two questions should be attempted.

One question from Section 1 (Questions 1, 2, 3) and
one question from Section 2 (Questions 4, 5, 6).

Write the numbers of the **two** questions you have attempted in the marks grid on the back cover of your answer booklet.

The value attached to each question is shown in the margin.

Credit will be given for appropriate models, diagrams, maps and graphs.

Marks may be deducted for bad spelling, bad punctuation and for writing that is difficult to read.

Note   The reference maps and diagrams in this paper have been printed in black only: no other colours have been used.

SCOTTISH QUALIFICATIONS AUTHORITY

Marks

SECTION 1

You must answer ONE question from this Section.

Question 1 (Rural Land Resources)

(a) Study Reference Map Q1.

Suggest why different National Parks attract widely differing numbers of visitors.   **5**

(b) *"Large numbers of visitors to a National Park can create environmental problems for those who live there."*

Referring to locations in a named National Park or upland area in the UK which you have studied:

  (i)   **give examples** of environmental problems which result from high numbers of visitors;

  (ii)   **describe** how such problems are tackled, and

  (iii)   **comment** on the effectiveness of these measures.   **10**

(c) The Lake District and Snowdonia are particularly noted for their glaciated scenery.

For **either** of these National Parks **or** another named glaciated upland area in the UK, **explain**, with the aid of annotated diagrams, how the main features of the physical landscape were formed.   **10**

    (25)

Reference Map Q1 (National Parks, visitor numbers, and main centres of population)

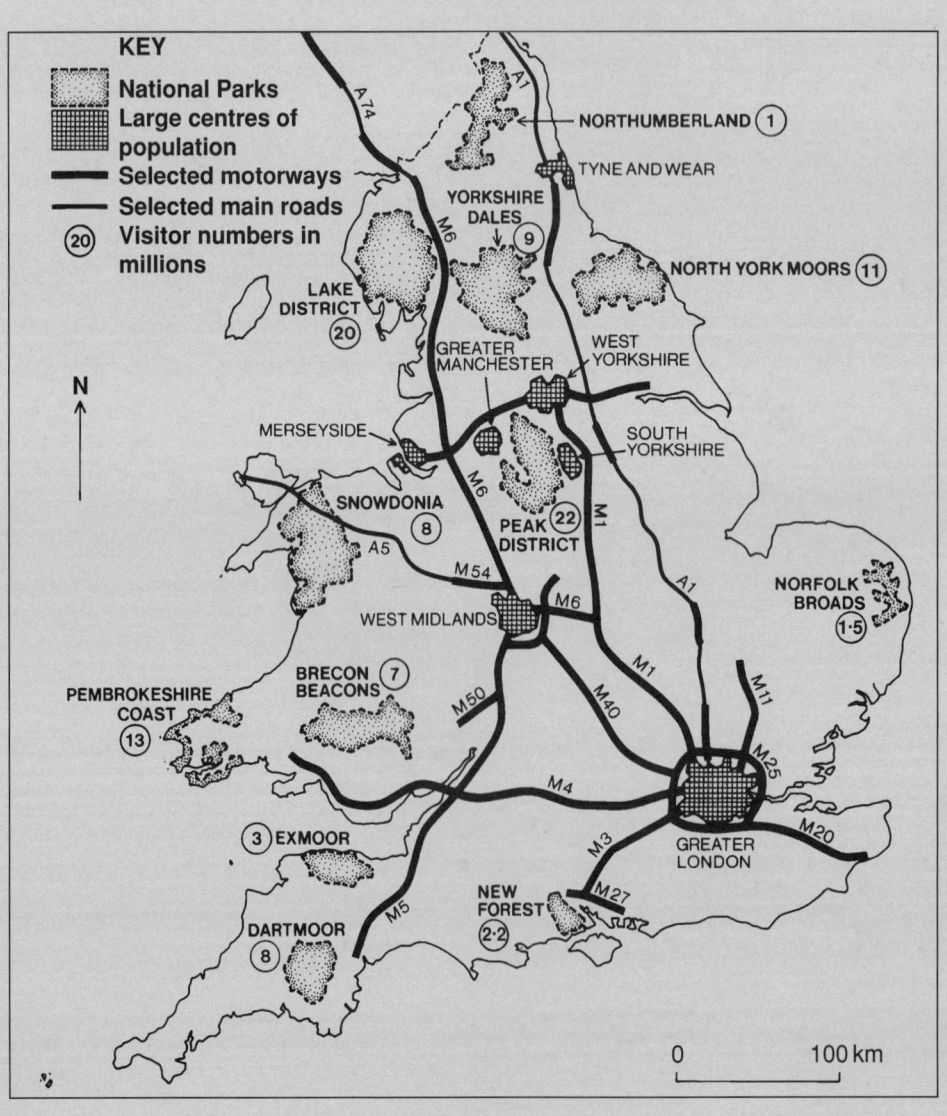

Marks

Question 2 (Rural Land Degradation)

"The rain clouds dropped a little spattering and hurried on to some other country. In the dust there were drop craters where the rain had fallen, and there were clean splashes on the corn and that was all."

(John Steinbeck)

(*a*) **Describe** the physical **and** human causes of land degradation in North America. **8**

(*b*) **Explain** the impact of land degradation on the people and environment of:
  (i)   North America, **and**
  (ii) **either** Africa north of the Equator **or** the Amazon Basin. **10**

(*c*) **Describe** the solutions which have been employed to control land degradation in **either** Africa north of the Equator **or** the Amazon Basin. **Comment** on the effectiveness of these methods. **7**

  (25)

[Turn over

Marks

Question 3 (River Basin Management)

(a) Study Reference Map Q3, Reference Diagram Q3A and Reference Diagram Q3B. Reference Diagram Q3B shows the variation in the discharge of the River Nile before and after the Aswan High Dam was built in 1963.

  (i) **Describe** and **account for** the pattern of river flow **before** the dam was built.

  (ii) **Describe** the ways in which the river's flow has changed since the completion of the dam.

7

(b) **Describe** and **account for** the social, economic and environmental benefits **and** adverse consequences of water control projects in a river basin you have studied in **either** Africa **or** North America.

13

(c) *"The next war in our region will be over the waters of the Nile."*

Dr Boutros Boutros-Ghali, Egyptian, former UN Secretary-General

For a river basin you have studied,

  (i) **explain** the political problems which have resulted, or may result, from the building of multi-purpose water projects, and

  (ii) **suggest** how these problems may be resolved.

5

(25)

Reference Map Q3 (The Nile Basin)

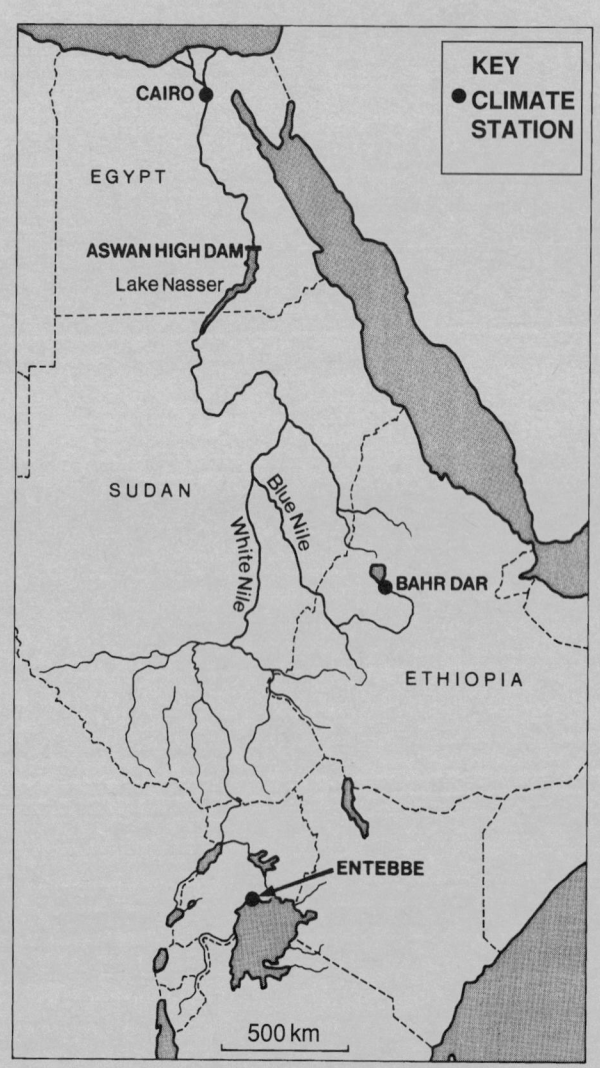

Page four

Question 3 – continued

Reference Diagram Q3A (Selected climate graphs)

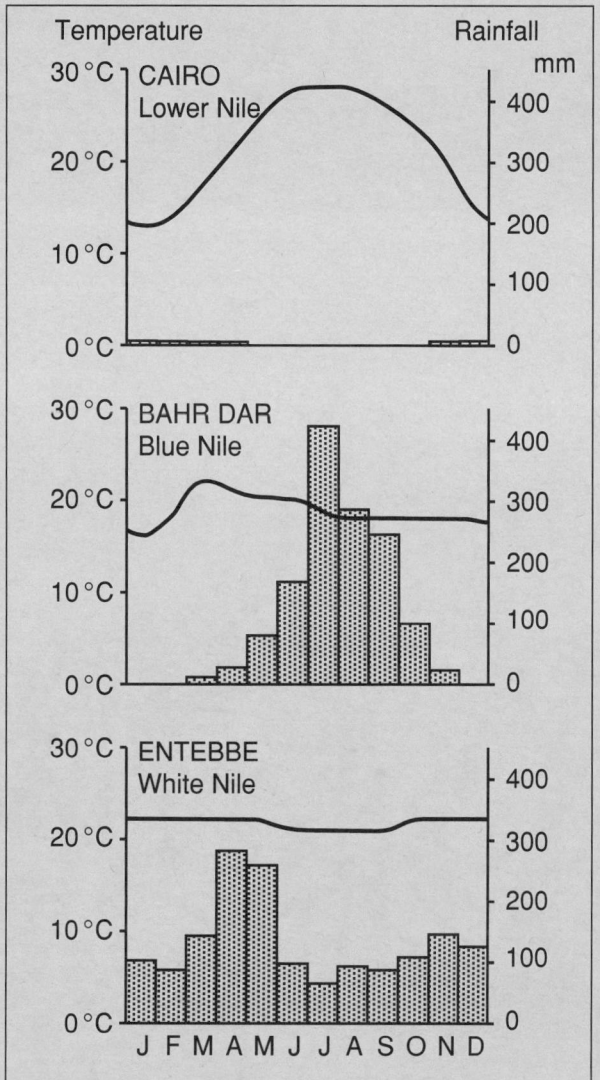

Reference Diagram Q3B (River Nile discharge before and after the building of the Aswan High Dam)

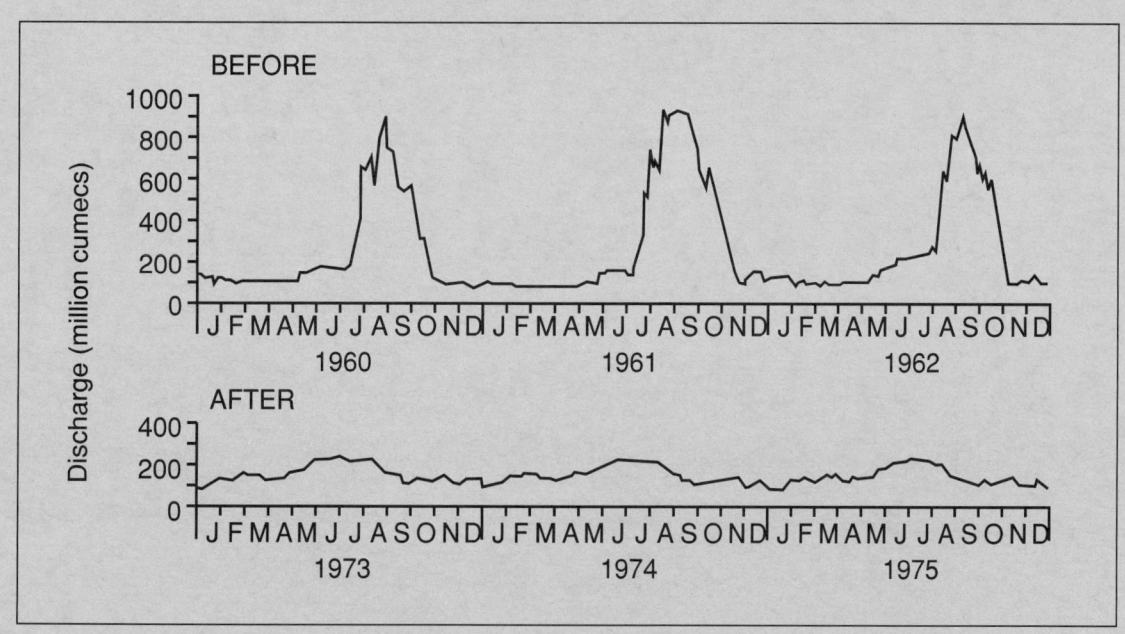

Marks

SECTION 2

You must answer ONE question from this Section.

Question 4 (Urban Change and its Management)

(*a*)   (i)   Study Reference Map Q4A.

With reference to the USA, **or** any other named country you have studied in the **Developed** World, **describe** and **explain** the distribution of major cities.   **5**

(ii)   Study Reference Map Q4B.

With reference to New York, **or** any named city you have studied in the **Developed** World, **describe** and **explain** the physical and human factors involved in its growth.   **5**

(iii)   In the second half of the 20th century, cities in the **Developed** World have undergone major changes in:

   • housing
   • industry
   • shopping
   • transport.

Choose **one** of the above and, for either New York **or** any named **Developed** World city you have studied, **describe** and **explain** the changes which have taken place.   **5**

(*b*)   Study Reference Diagram Q4.

Referring to Mexico City **or** a named city you have studied in the **Developing** World,

   (i)   **describe** and **account** for its rapid growth, and

   (ii)   **outline** the social and environmental problems which have resulted from this growth.   **10**

(25)

Reference Map Q4A (Distribution of major cities in the USA)

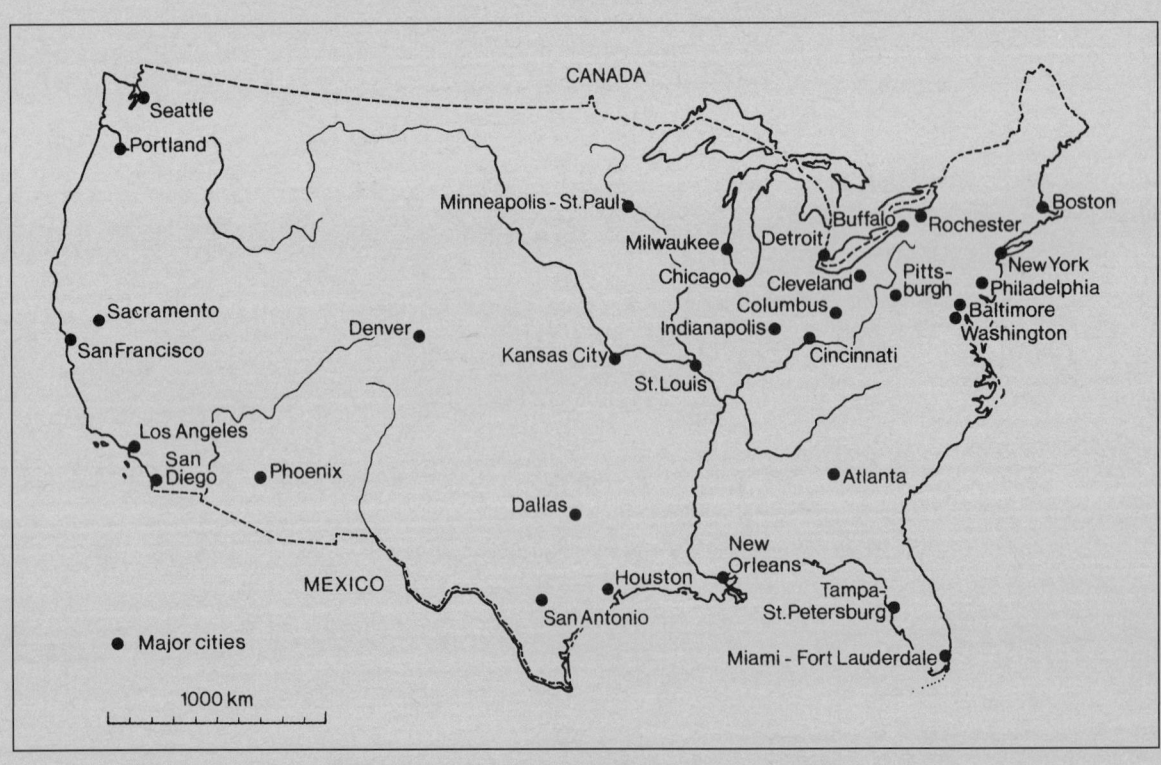

Question 4 – continued

Reference Map Q4B (The location of New York City)

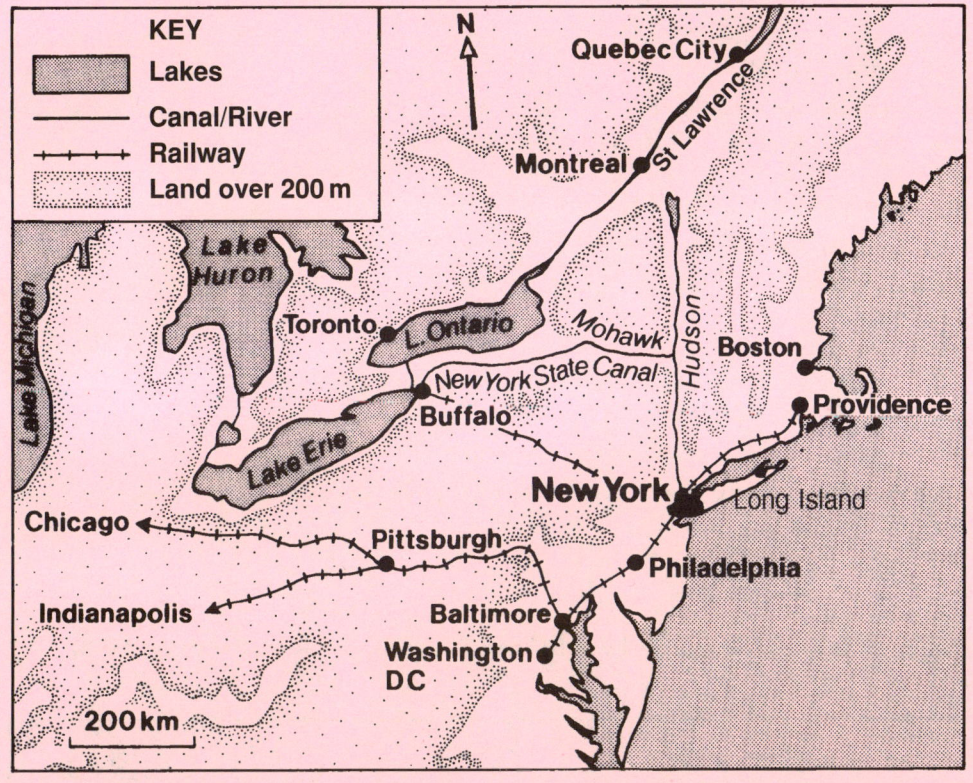

Reference Diagram Q4 (Population growth in Mexico City)

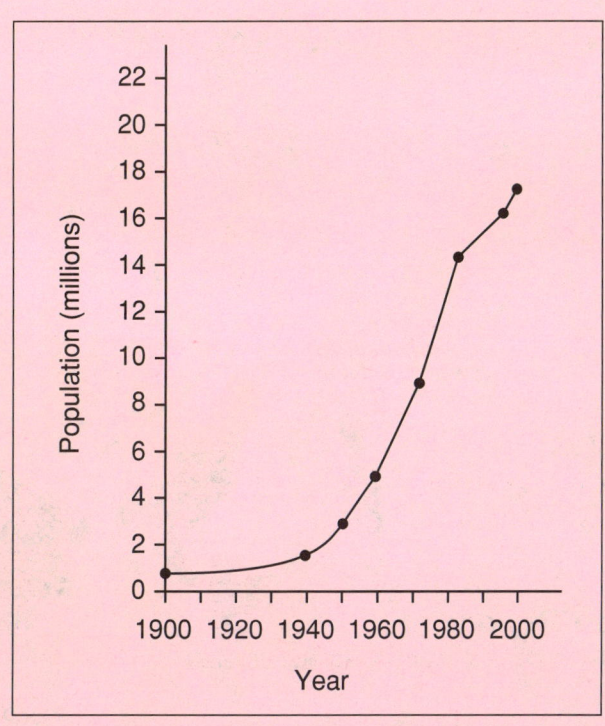

Marks

Question 5 (European Regional Inequalities)

(a) Study Reference Map Q5A which shows the areas which receive Objective 1 support. These areas are defined by the European Union as "lagging behind".

   (i) **Describe** the distribution of the areas classified by the European Union as "lagging behind".   **4**

   (ii) **Describe** the benefits which an Objective 1 area might receive from the European Union to assist in its development.   **4**

(b) Study Reference Map Q5B and Reference Table Q5.

   "Many European countries suffer from regional inequalities."

   To what extent does the information shown illustrate the existence of regional inequalities in France?   **6**

(c) For any **named** country you have studied in the European Union,

   (i) **describe** the human and physical factors which have contributed to the development of regional inequalities, and   **7**

   (ii) **describe** the steps taken by the national government to reduce these inequalities, and comment on their effectiveness.   **4**

          (25)

Reference Map Q5A (Regions of the EU receiving Objective 1 funding, 1998)

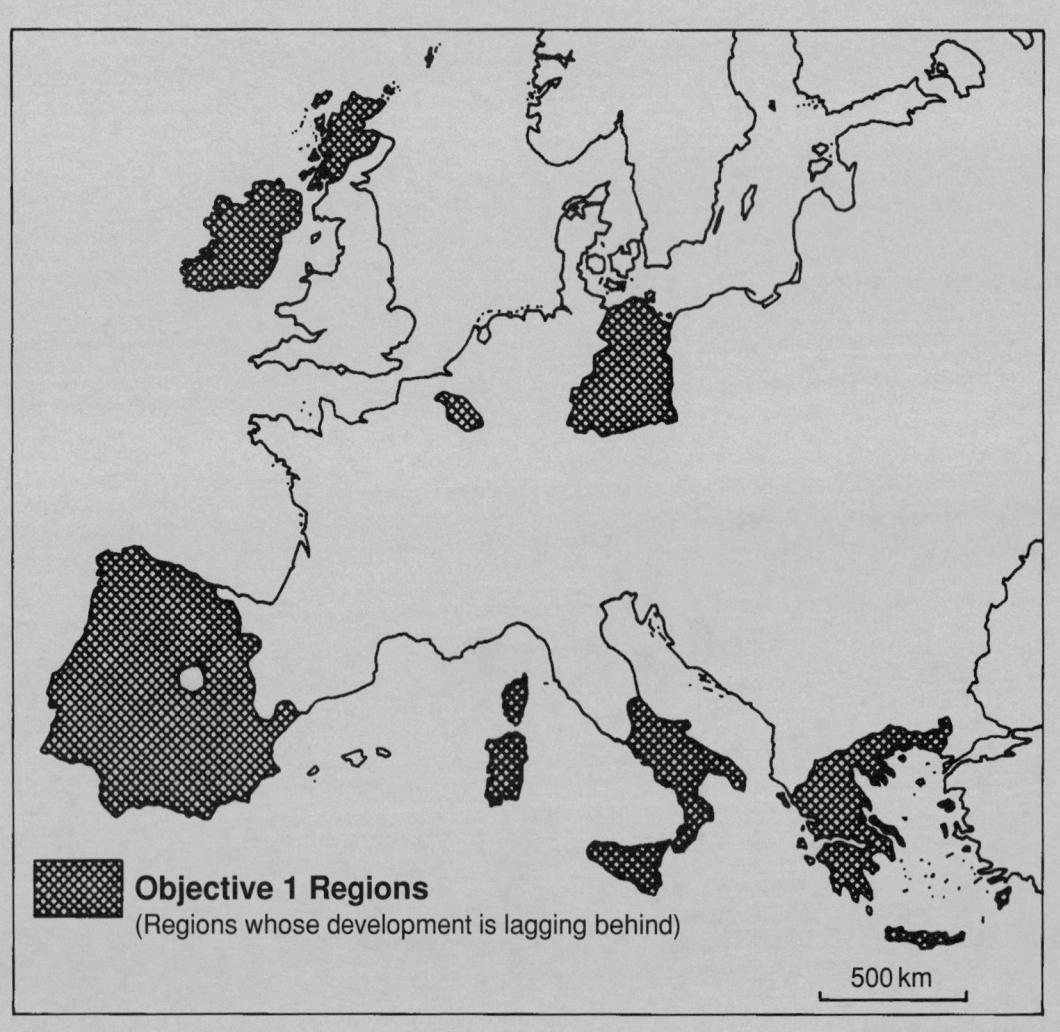

Objective 1 Regions
(Regions whose development is lagging behind)

500 km

Question 5 – continued

Reference Map Q5B (Regions of France)

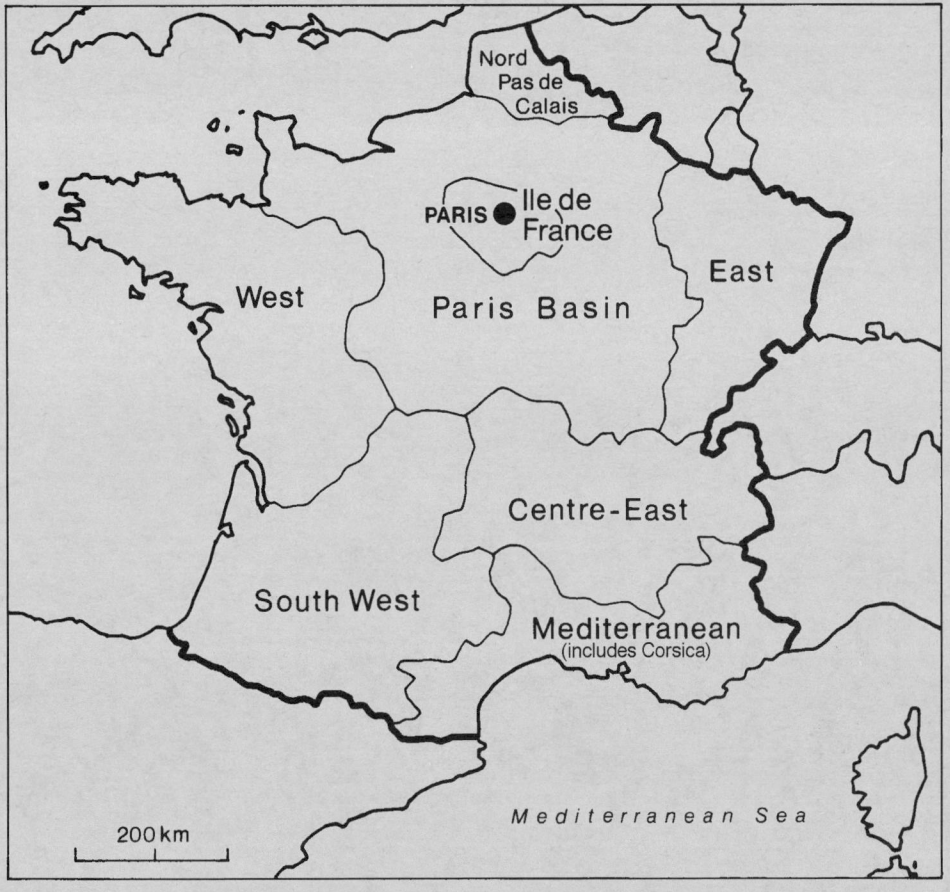

Reference Table Q5 (Selected socio-economic data for Regions of France)

Region	Population Density (per sq km)	Migration Rate per thousand	GDP per capita (ECU)	Unemployment (percentage)
Nord/Pas de Calais	320·5	−2·7	91	16·3
Ile de France (includes Paris)	904·2	−0·8	173	10·9
Paris Basin	71·2	0·7	105	13·3
East	105·1	−2·4	106	9·7
West	88·6	4·0	95	11·6
South West	58·1	4·0	99	12·3
Centre-East	97·2	1·5	109	11·4
Mediterranean	100·5	5·5	98	15·9

Marks

Question 6 (Development and Health)

(a) Study Reference Table Q6.

The table shows that there are considerable differences in levels of development between countries in the **Developing** World. Referring to these countries and/or others in the Developing World which you have studied, **suggest reasons** why such differences **between** countries exist.

6

(b) Study Reference Diagram Q6.

With reference to countries of the Developing World with which you are familiar, **describe** and **suggest reasons** for the differences in the provision of social services between urban and rural areas.

5

(c) Primary Health Care strategies have been introduced by many countries in the Developing World in an effort to improve the health of the population.

Give examples of Primary Health Care strategies and **comment** on their effectiveness in improving health and controlling disease in areas you have studied.

6

(d) With reference to malaria **or** another water-related disease, **describe** the measures used to control the disease.

8

(25)

Reference Table Q6 (Indicators of development for selected countries)

Indicator	Saudi Arabia	South Korea	Ethiopia
GNP per capita (US dollars)	6910	8600	100
Life expectancy (years)	70	74	46
Infant mortality rate (per 1000 live births)	46	11	116
Birth rate (per 1000)	35	14	45

Reference Diagram Q6 (Access to social services in the Developing World)

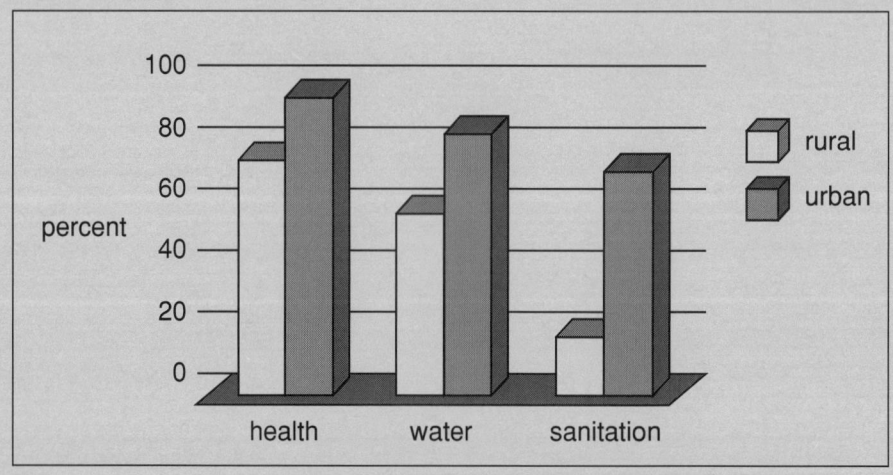

[END OF QUESTION PAPER]

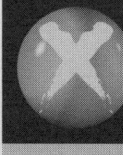

[BLANK PAGE]

X042/301

NATIONAL QUALIFICATIONS 2003	MONDAY, 2 JUNE 9.00 AM – 10.30 AM	GEOGRAPHY HIGHER Core

Attempt **all** questions.

The value attached to each question is shown in the margin.

Credit will be given for appropriate models, diagrams, maps and graphs.

Marks may be deducted for bad spelling, bad punctuation and for writing that is difficult to read.

Note   The reference maps and diagrams in this paper have been printed in black only: no other colours have been used.

SCOTTISH
QUALIFICATIONS
AUTHORITY

Landranger Series

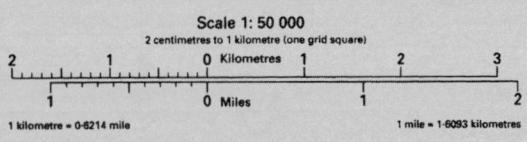

Scale 1: 50 000

2 centimetres to 1 kilometre (one grid square)

```
2          1          0  Kilometres  1          2          3
|__|__|__|__|__|__|__|__|__|__|__|__|__|__|__|__|__|__|__|

  1              0  Miles          1                    2
|__|__|__|__|__|__|__|__|__|__|__|__|__|__|__|__|
```

1 kilometre = 0·6214 mile   1 mile = 1·6093 kilometres

Extract No 1326/117

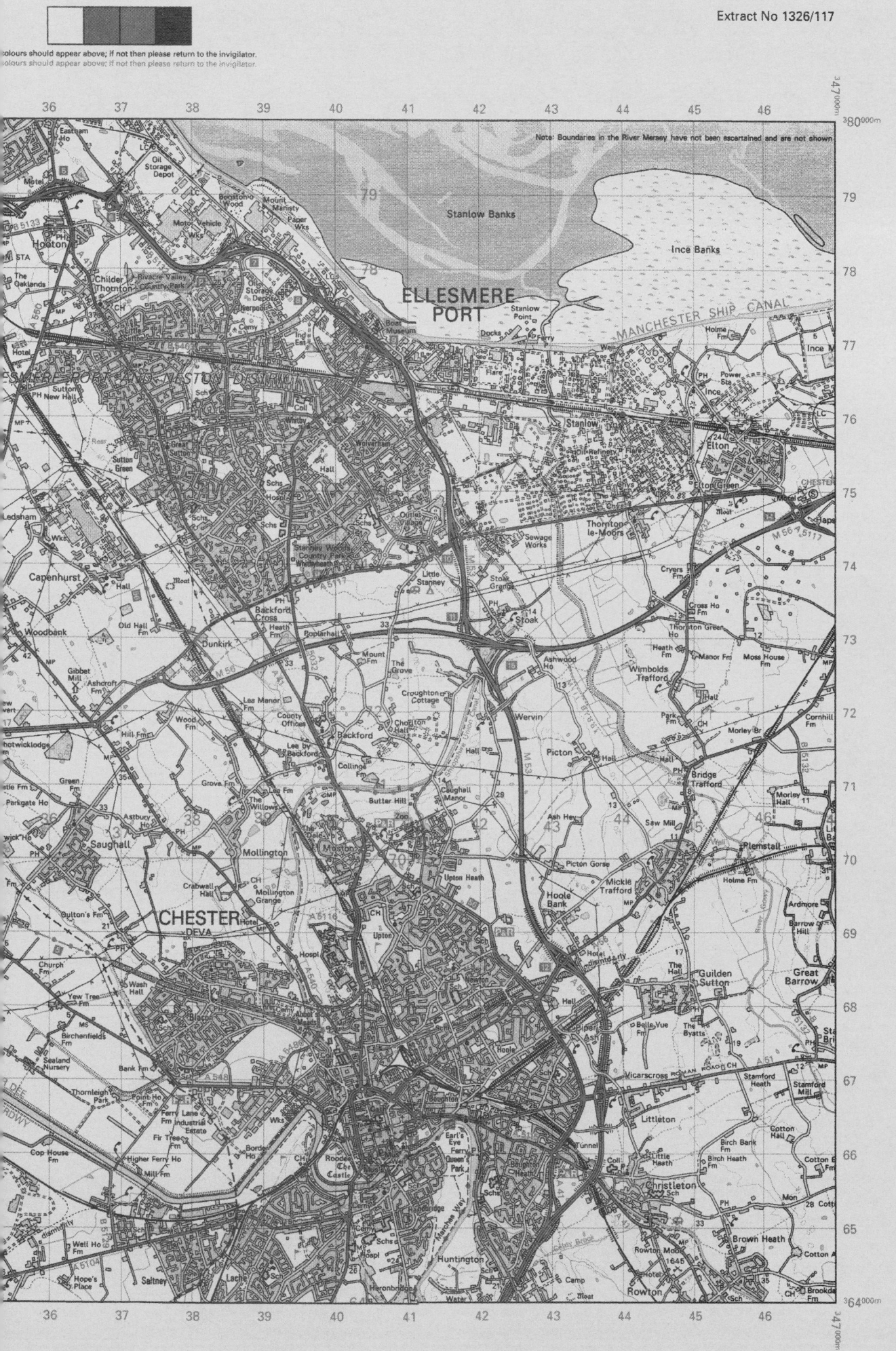

Marks

Question 1

Study Reference Map Q1.

Describe and **account for** the variation in rainfall within West Africa.

7

Reference Map Q1 (Rainfall across West Africa)

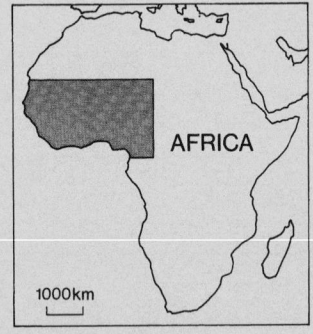

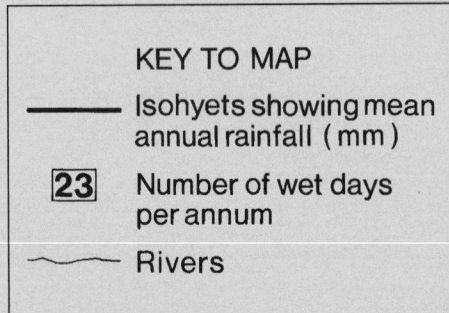

KEY TO MAP

—— Isohyets showing mean annual rainfall (mm)

23 Number of wet days per annum

~~~ Rivers

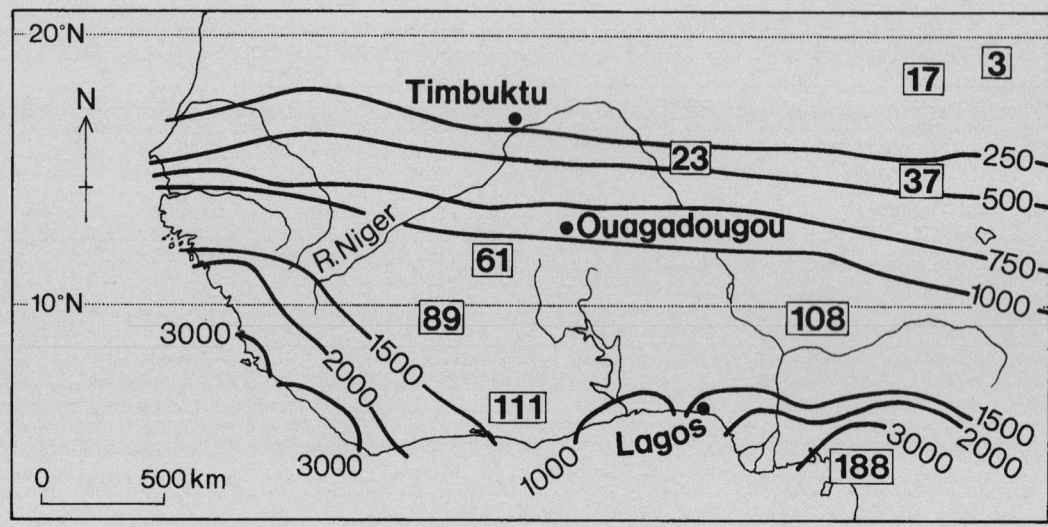

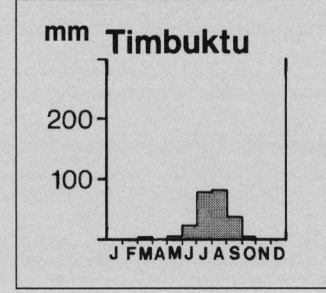

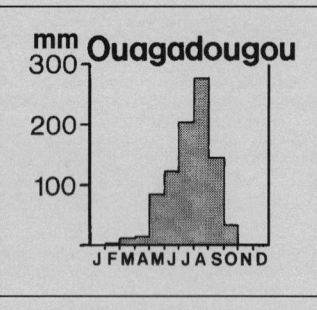

Marks

Question 2

Study OS map extract number 1326/117: Chester (*separate item*).

(*a*) Examine the course of the River Gowy between 467670 and its mouth at 431775.

Describe the evidence which would suggest that the river in this section is in its **lower** course.　　　　3

(*b*) Waterfalls are frequently found in the **upper** course of a river valley.

Describe fully how a waterfall may be formed.　　　　3

[Turn over

Marks

Question 3

ATTEMPT EITHER QUESTION 3A OR QUESTION 3B

3A

Reference Diagrams Q3A, B, C and D show different types of mass movement on slopes.

Choose any **two** of the mass movements shown, and, for each, **describe** and **explain** the conditions and processes which encourage it to take place.

6

Reference Diagram Q3 (Selected mass movements on slopes)

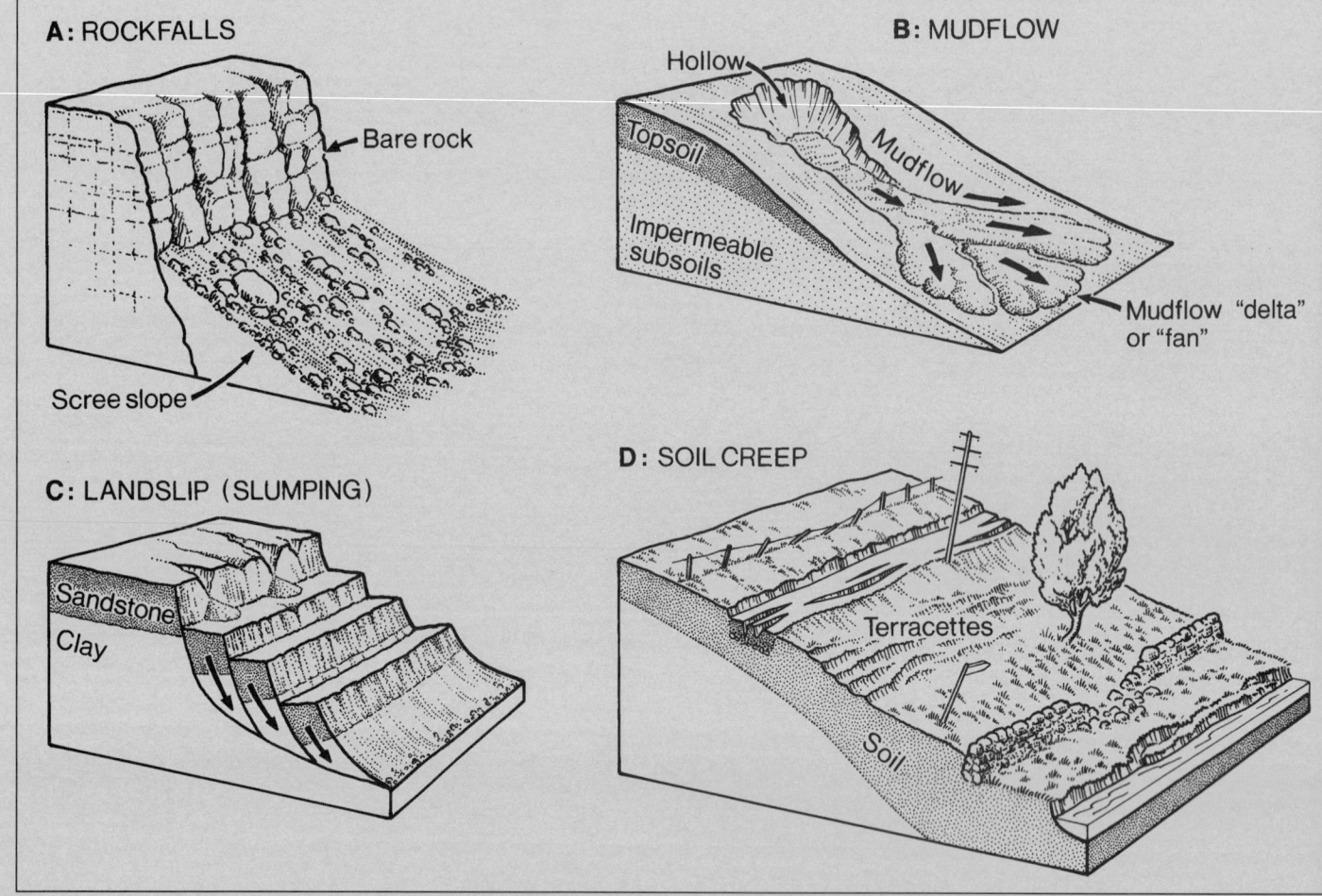

Marks

Question 3—continued

OR

3B

Study Reference Diagram Q3E.

Describe and **explain** the processes which have created the **physical** landscape shown in the diagram.

6

Reference Diagram Q3E (A scarp and vale landscape in England)

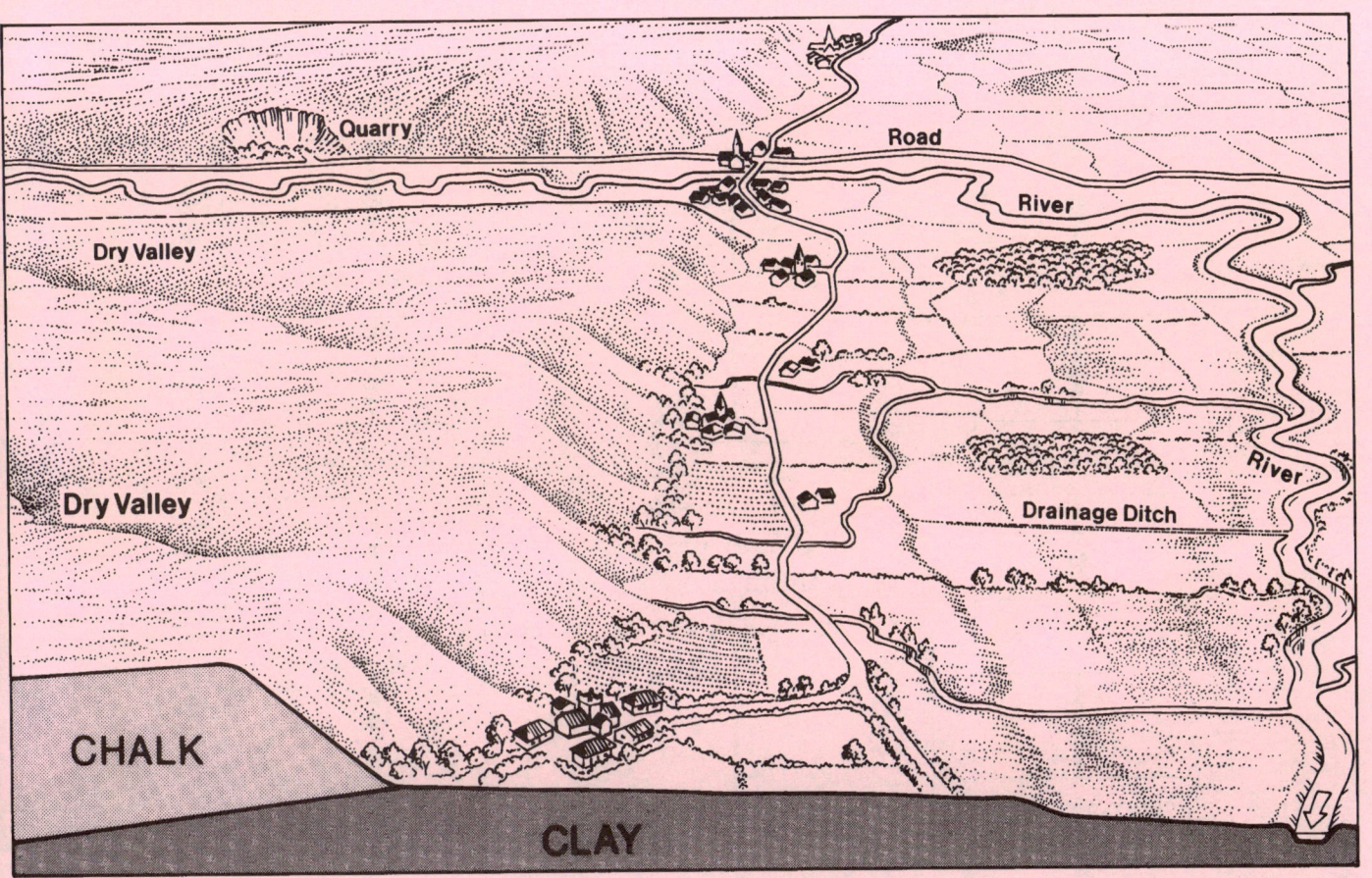

[Turn over

Marks

Question 4

Study Reference Diagram Q4.

Plant succession describes the changes in vegetation that develop through time in a particular habitat.

Describe and **explain** the plant succession for **either** a derelict urban habitat **or** a sand dune habitat. You should make reference to specific plants.

6

Reference Diagram Q4 (Plant succession in two selected habitats)

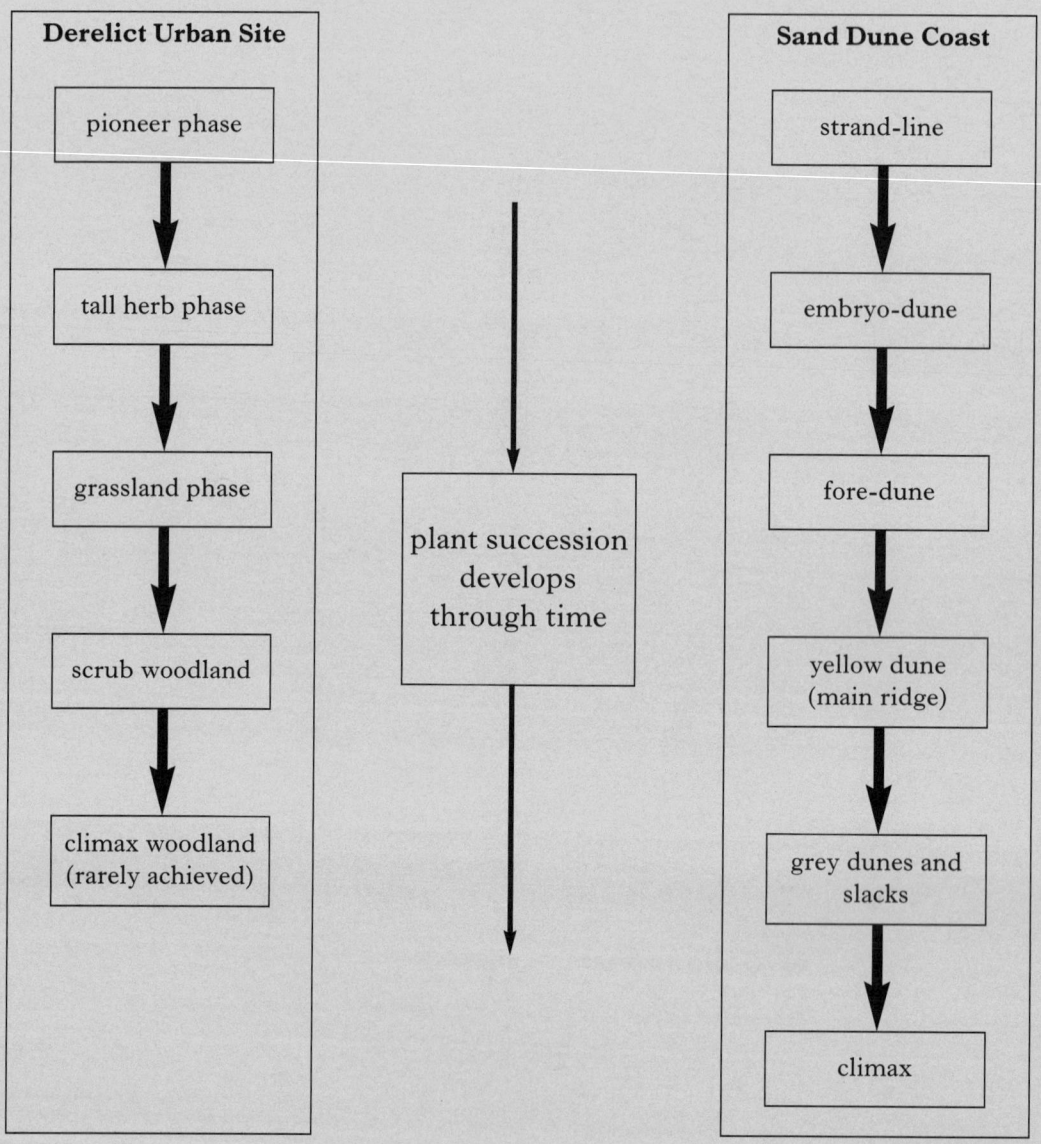

Marks

Question 5

Study Reference Diagrams Q5A and Q5B, which show population change over time.

With reference to **either** a named Developed Country (EMDC*) **or** a named Developing Country (ELDC*), **discuss** the factors which influenced the changes in the Birth Rate and the Death Rate during the last century.

6

Reference Diagram Q5A
(A developed country or EMDC*)

Reference Diagram Q5B
(A developing country or ELDC*)

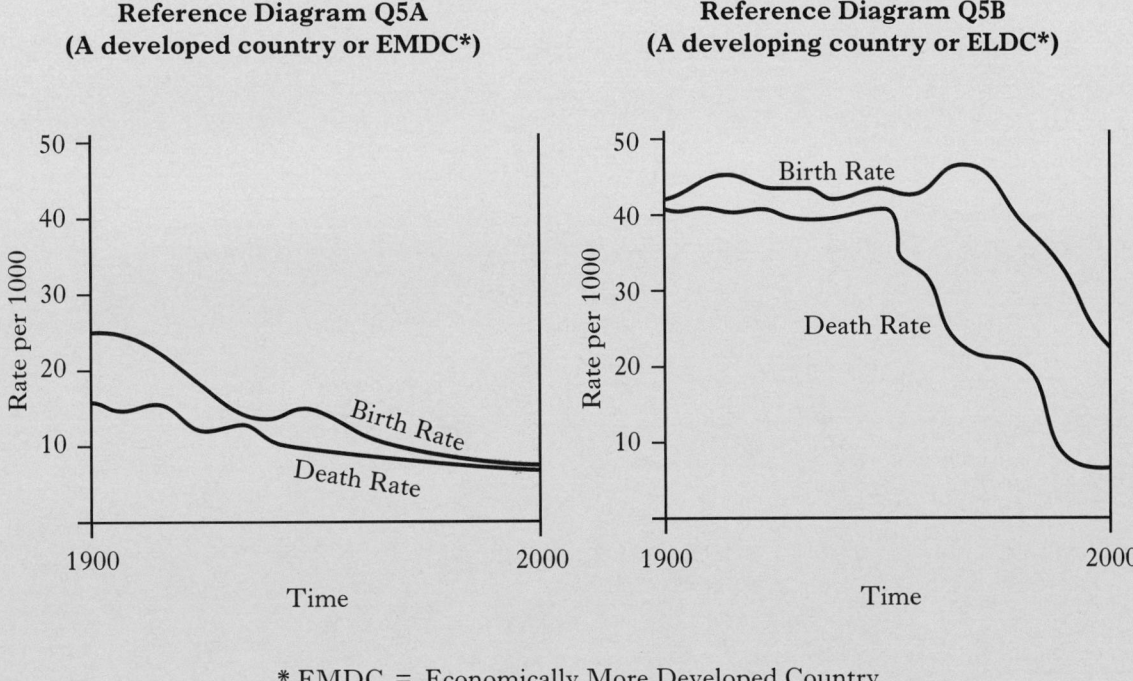

* EMDC = Economically More Developed Country
* ELDC = Economically Less Developed Country

[Turn over

Marks

Question 6

Study Photographs Q6A and Q6B.

"Both extensive commercial agriculture and intensive peasant farming continue to change and adopt new practices."

For **either** extensive commercial agriculture **or** intensive peasant farming, **describe** how such changes and new practices have affected the people and the farming landscape in a named location you have studied.

6

Photograph Q6A (Extensive commercial agriculture—harvesting, North America)

Photograph Q6B (Intensive peasant farming—ploughing, South East Asia)

Marks

Question 7

Study OS map extract number 1326/117: Chester (*separate item*), and Reference Maps Q7A and Q7B.

The area shown in the OS map extract has a large number of industrial complexes.

Explain fully the factors that make the map area attractive for the location of industry. In your answer you should quote map evidence and give examples from the industrial areas shown on Reference Map Q7B.

6

Reference Map Q7A (Location of the OS map extract)

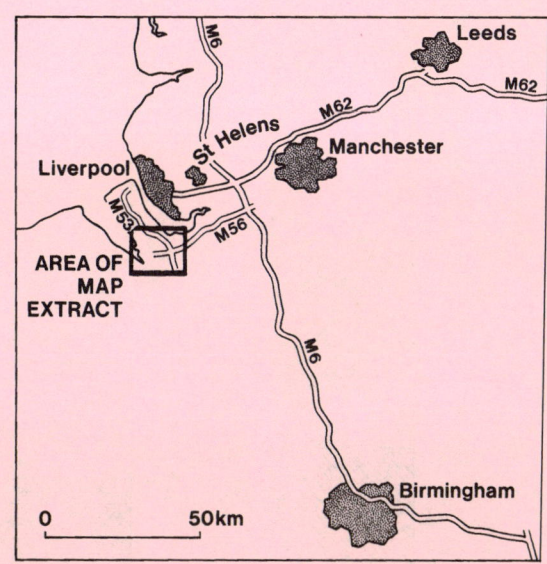

Reference Map Q7B (Location of the main industrial areas)

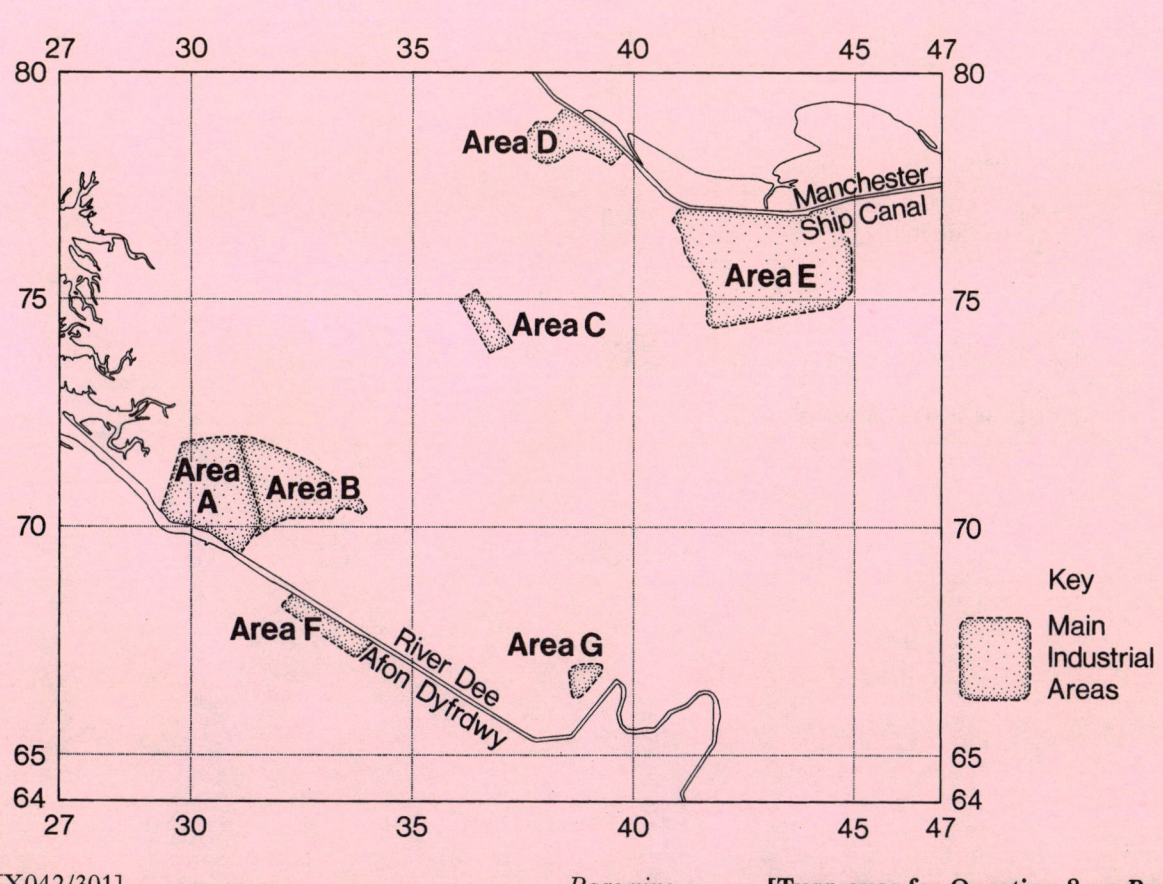

Page nine **[Turn over for Question 8 on *Page ten***

Marks

Question 8

Study Reference Diagram Q8.

The diagram shows selected changes in land use from the Central Business District to the edge of a typical city in the **Developed** World.

With reference to the city you have studied, **describe** and **explain** the differences in land use from the centre of the city to the edge.

7

Reference Diagram Q8 (Land use transect from the CBD to the edge of a city)

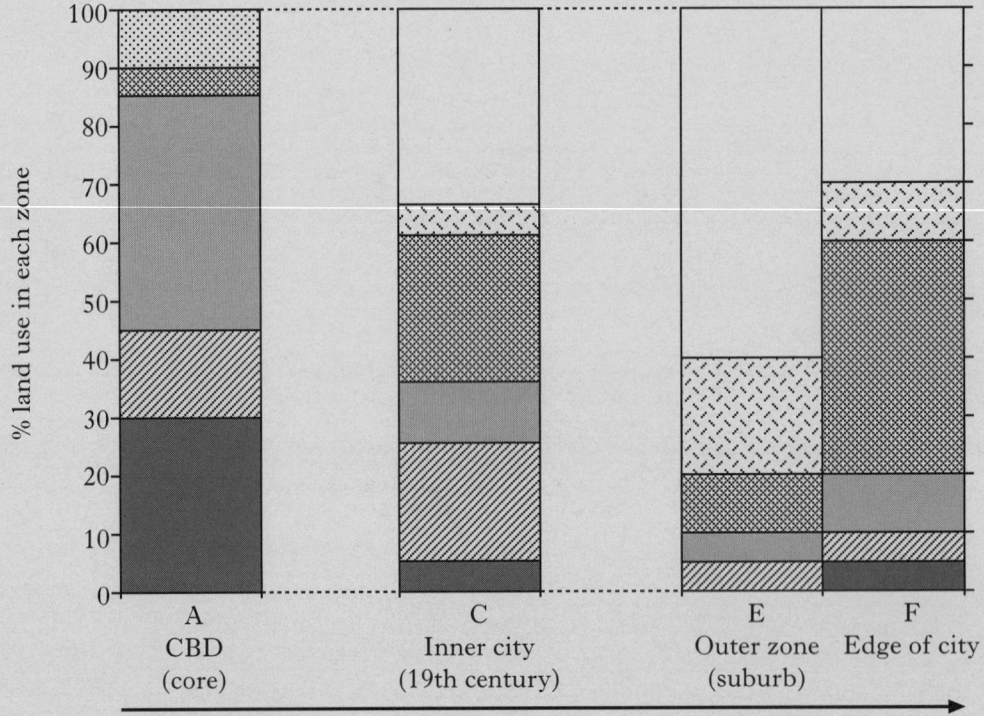

Increasing distance from the city centre

Land use

	residential		offices
	open space		convenience shops, eg newsagent
	public buildings		comparison shops, eg electrical shop
	industry and warehouses		

[END OF QUESTION PAPER]

X042/303

NATIONAL
QUALIFICATIONS
2003

MONDAY, 2 JUNE
10.50 AM – 12.05 PM

GEOGRAPHY
HIGHER
Applications

Two questions should be attempted.

One question from Section 1 (Questions 1, 2, 3) and
one question from Section 2 (Questions 4, 5, 6).

Write the numbers of the **two** questions you have attempted in the marks grid on the back cover of your answer booklet.

The value attached to each question is shown in the margin.

Credit will be given for appropriate models, diagrams, maps and graphs.

Marks may be deducted for bad spelling, bad punctuation and for writing that is difficult to read.

Note The reference maps and diagrams in this paper have been printed in black only: no other colours have been used.

SCOTTISH
QUALIFICATIONS
AUTHORITY

Marks

SECTION 1

You must answer ONE question from this Section.

Question 1 (Rural Land Resources)

(a) Study Reference Map Q1.

"*Ingleborough Hill is one of the famous Three Peaks in the Yorkshire Dales National Park. Over millions of years natural processes have created an exceptional Carboniferous Limestone landscape.*"

With the aid of annotated diagrams, **describe** and **explain** how the main features of a Carboniferous Limestone landscape were formed. Both underground and surface features should be mentioned.

10

(b) Upland landscapes provide economic and social opportunities for a variety of land uses, for example

> agriculture
>
> mineral exploitation
>
> forestry
>
> recreation and tourism
>
> industry
>
> water storage.

For the Yorkshire Dales, **or** any other upland area you have studied, **explain two** of the opportunities listed above.

6

(c) The area around Ingleton and Ingleborough Hill is described as a "honeypot", as it is a very popular area for tourists to visit.

(i) **Explain** the problems and conflicts which may arise in and around any **named** "honeypot" area you have studied.

(ii) **Suggest** ways in which these problems may be resolved.

9

(25)

Reference Map Q1 (Ingleborough Hill and part of the Yorkshire Dales National Park)

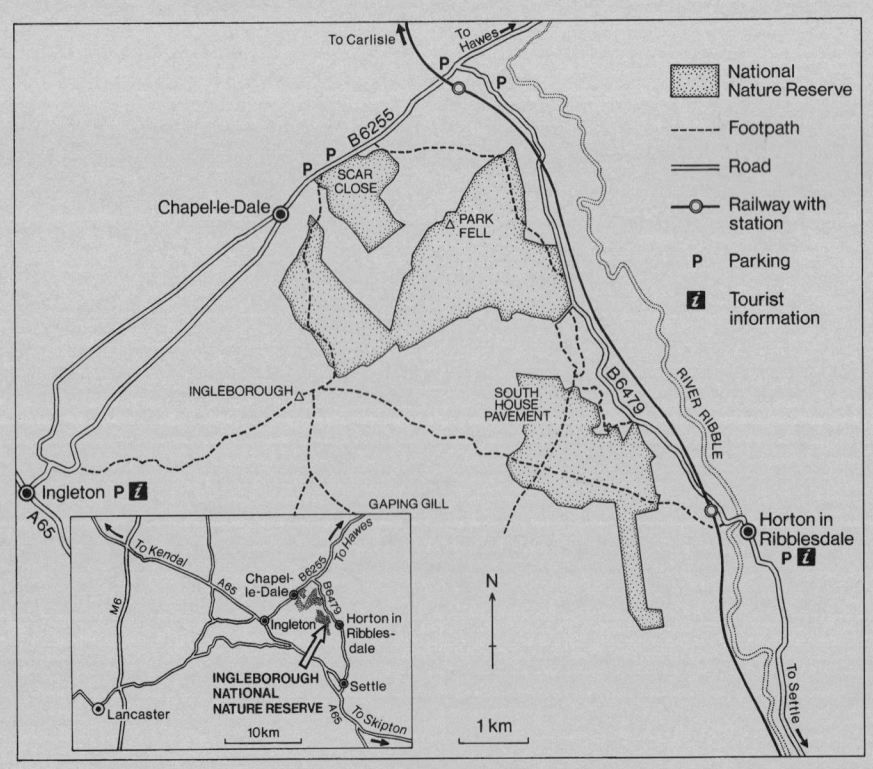

Marks

Question 2 (Rural Land Degradation)

(a) Study Reference Table Q2.

Select **three** of the factors listed and, referring to areas you have studied in **either** Africa north of the Equator **or** the Amazon Basin,

    (i) **explain** how each factor can lead to land degradation, and   7

    (ii) **describe** the impact of land degradation on the people and the environment.   8

(b) Referring to named areas of North America **and either** Africa north of the Equator **or** the Amazon Basin,

    (i) **describe** some of the measures which have been taken to conserve soil and limit land degradation, and

    (ii) **comment** on the effectiveness of the measures.   10

                                                           (25)

Reference Table Q2 (Selected factors which can lead to land degradation)

Climatic variability

Deforestation

Overcultivation

Overgrazing

[Turn over

Marks

Question 3 (River Basin Management)

(*a*)   Study Reference Maps Q3A and Q3B.

For **either** Africa **or** North America, **describe** and **explain** the general distribution of river basins.

5

(*b*)   Study Reference Maps Q3C and Q3D.

For the Revelstoke Dam and Reservoir Scheme on the Columbia River, **or** a similar water control project you have studied in **either** Africa **or** North America,

(i)   **explain** the physical factors which have to be considered when selecting sites for dams and their associated reservoirs,

5

(ii)   **explain** the project's effects on the hydrological cycle of the river basin, and

5

(iii)   **explain** the social and economic benefits **and** adverse consequences of the project.

10

(25)

Reference Map Q3A
(Major river basins of Africa)

Reference Map Q3B
(Major river basins of North America)

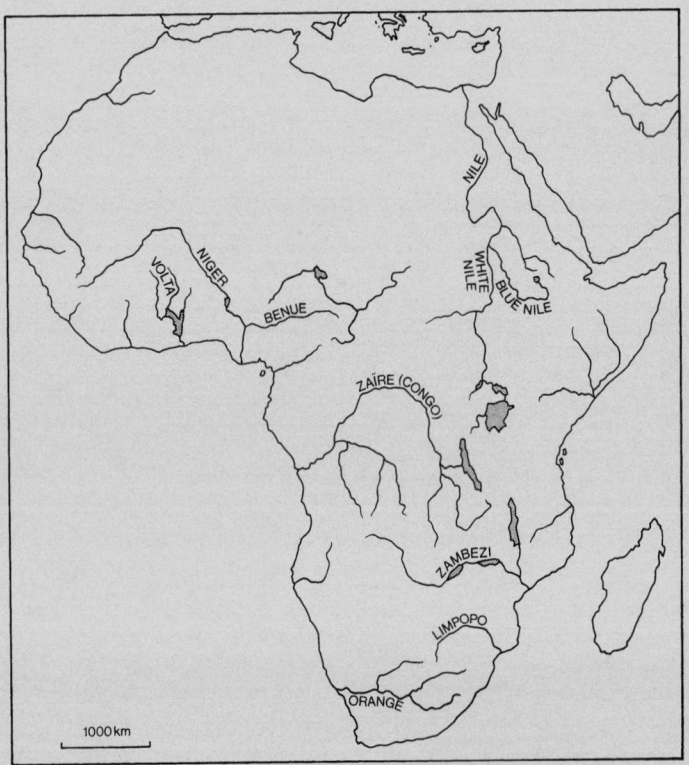

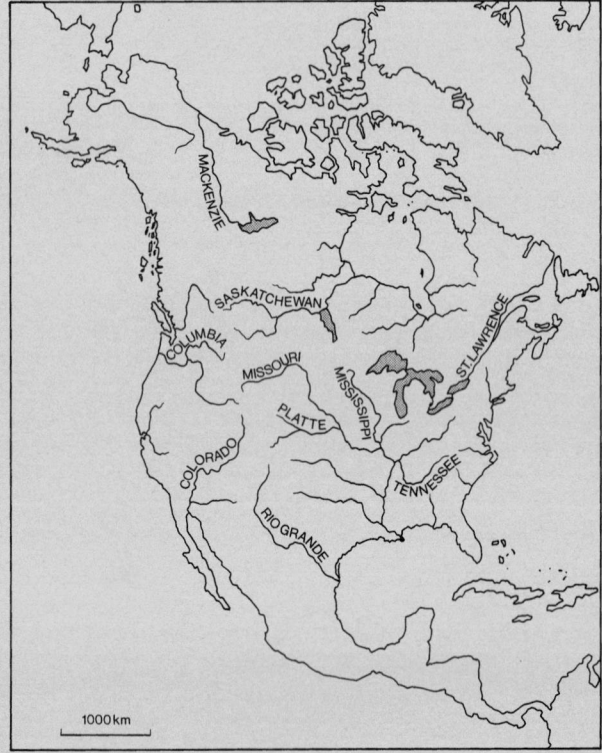

Question 3 – continued

Reference Map Q3C (Columbia River Basin and location of Revelstoke)

Reference Map Q3D (Revelstoke Reservoir)

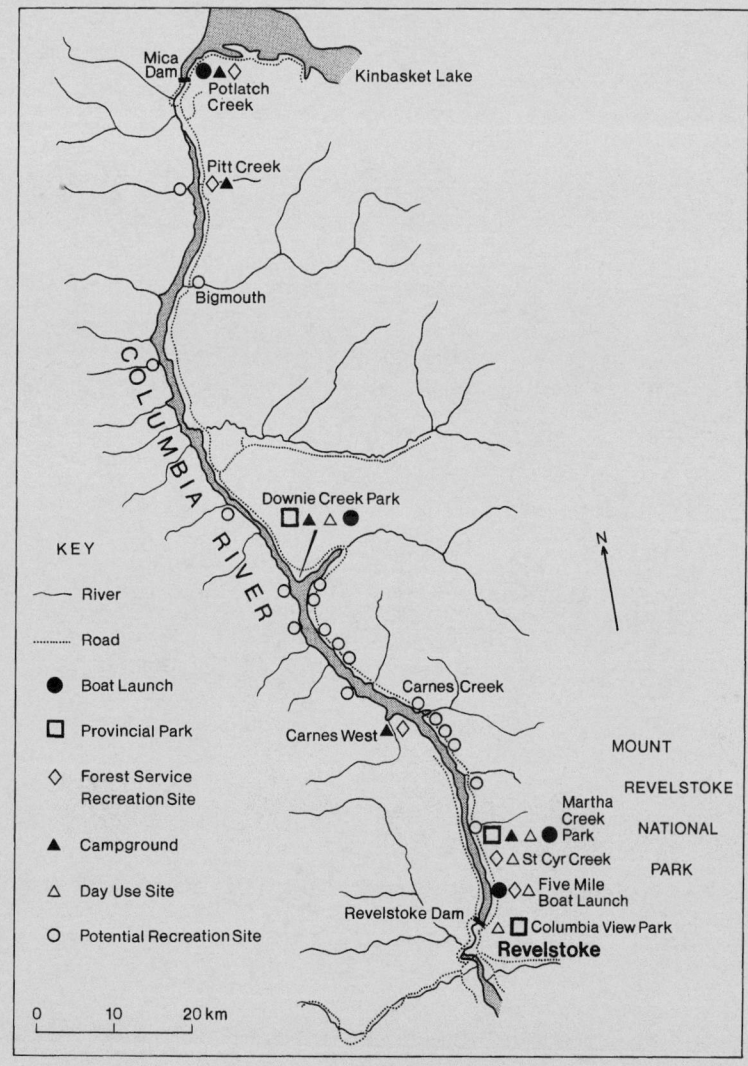

SECTION 2

You must answer ONE question from this Section.

Question 4 (Urban Change and its Management)

(a)   Study Reference Diagram Q4.

   (i)   **Describe** the trends in urban population shown in the diagram.

   (ii)   Referring to cities you have studied, **explain** the differing growth rates between cities in the Developed and the Developing World.   **7**

(b)   Study Reference Table Q4.

   The table shows the size of the problem facing planners in the cities of the **Developing** World.

   For any city you have studied in the **Developing** World,

   (i)   **describe** the problems likely to be faced by urban slum dwellers, and

   (ii)   **identify** strategies which have been used to improve the lives of the slum dwellers and **comment** on their effectiveness.   **10**

(c)   Study Reference Map Q4.

   For the Salford Quays dockland redevelopment area of Manchester, **or** a specific redevelopment area of a **named** city in the **Developed** World,

   (i)   **explain** why the redevelopment was considered necessary, and

   (ii)   **describe** the changes that have taken place and **assess** their effectiveness.   **8**

   (25)

Reference Diagram Q4 (Urban population by region 2000 and 2030, projected)

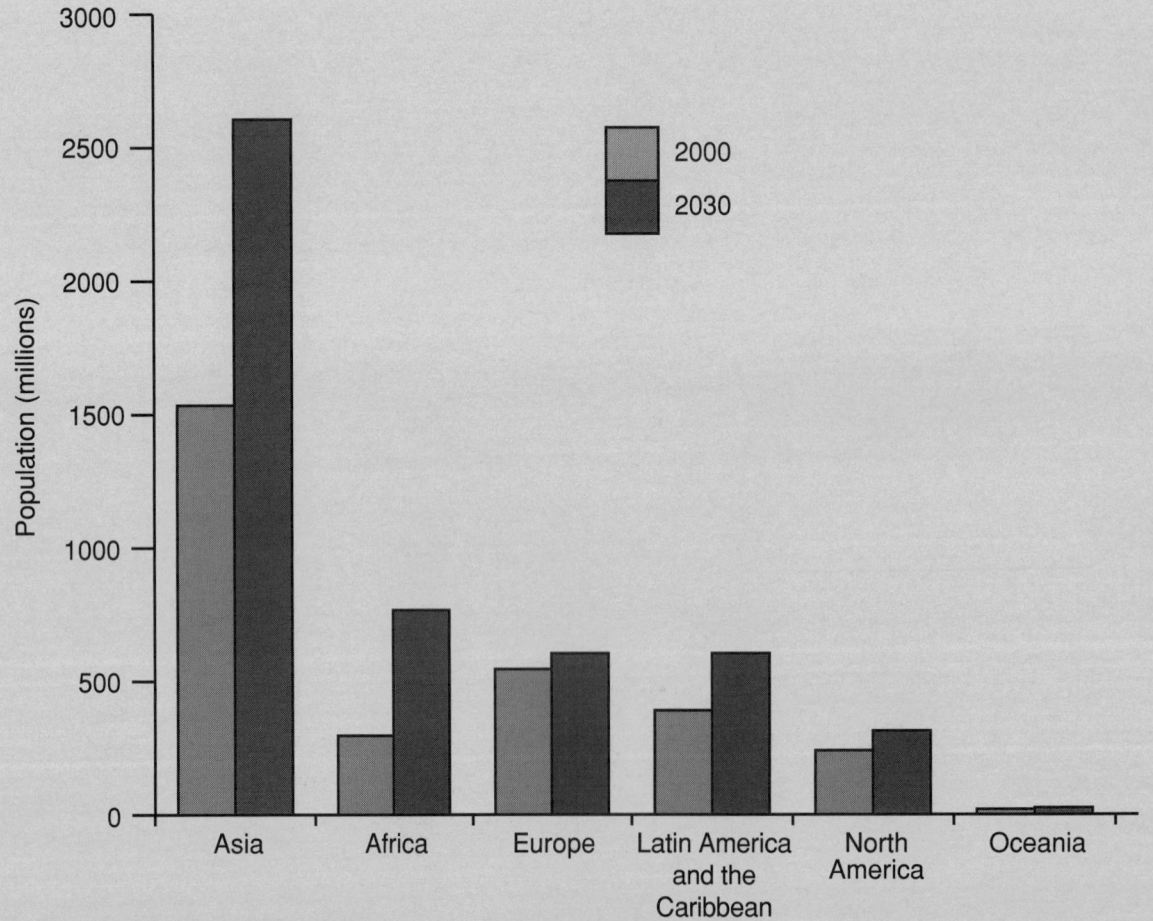

Question 4 – continued

Reference Table Q4 (Slum population in major Indian cities)

Name	Total Population	Slum Population	% Slum Dwellers
Greater Mumbai (Bombay)	11 914 398	5 823 510	48%
Delhi	9 817 439	1 854 685	18%
Kolkata (Calcutta)	4 580 544	1 490 811	32%
Bangalore	4 292 223	345 200	8%
Chennai (Madras)	4 216 268	1 079 414	25%

Reference Map Q4 (Salford Quays, Dockland Redevelopment, Manchester)

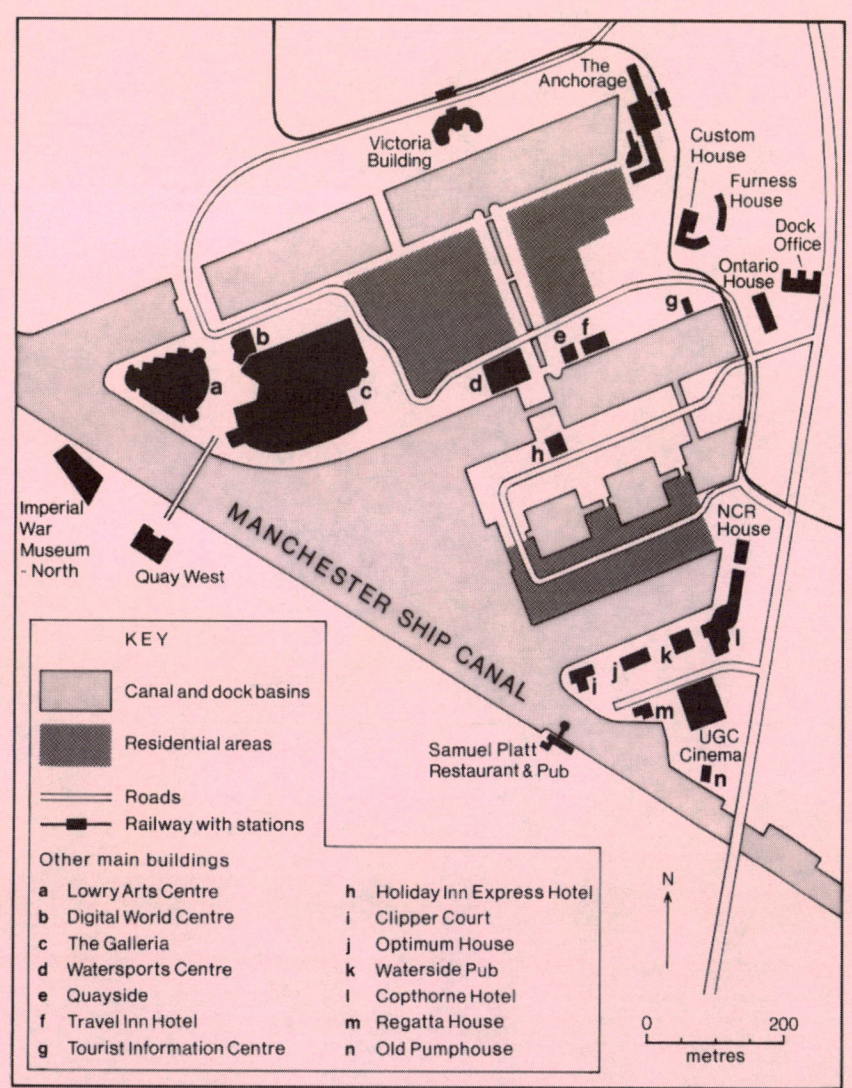

Question 5 (European Regional Inequalities)

(a) Study Reference Maps Q5A to Q5E.

   (i) **To what extent** does this information suggest that there were regional inequalities in the UK in 2000?

             6

   (ii) **Describe** the physical **and** human factors which have led to regional inequalities in the UK.

             10

(b) For **either** the UK **or** any other country you have studied in the European Union,

   (i) **describe** the steps taken by the European Union and the national government to reduce regional inequalities, and

   (ii) **comment** on the effectiveness of these measures.

             9

             (25)

Reference Map Q5A (UK Statistical Regions)

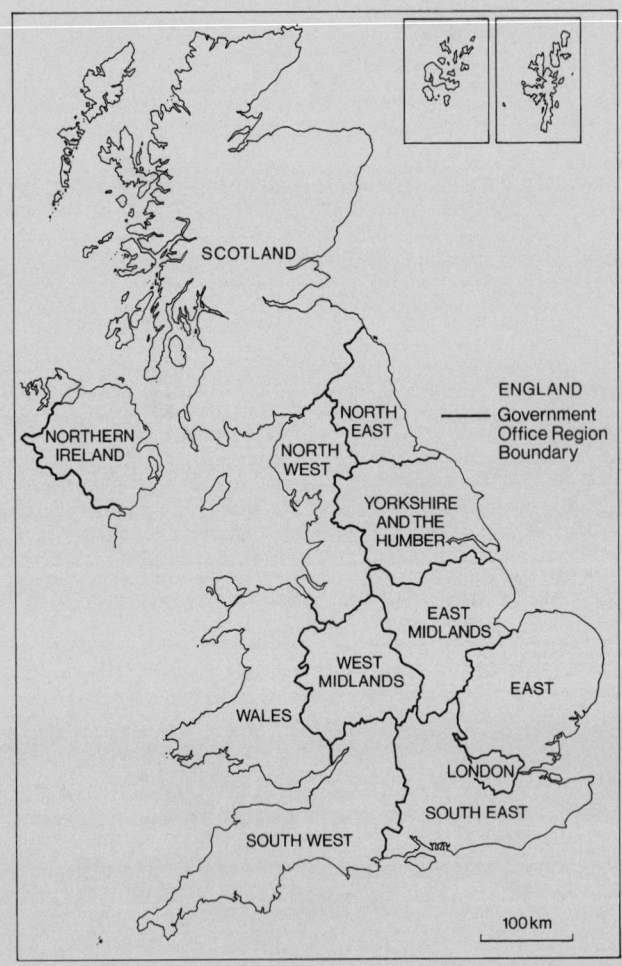

Question 5 – continued

Reference Map Q5B
(Average house prices, January 2000)

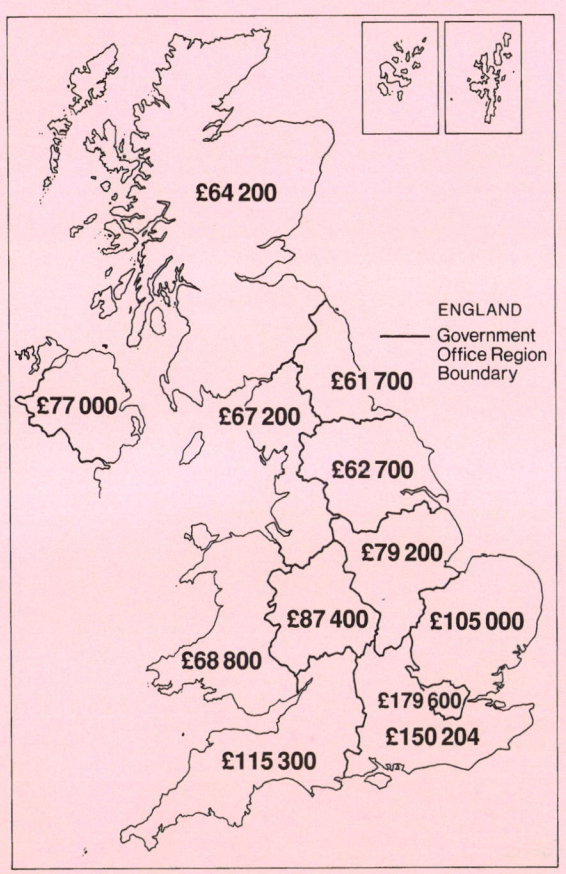

ENGLAND
Government
Office Region
Boundary

£64 200

£77 000

£61 700

£67 200

£62 700

£79 200

£87 400

£105 000

£68 800

£179 600

£150 204

£115 300

Reference Map Q5C
(Unemployment rate, Spring 2000)

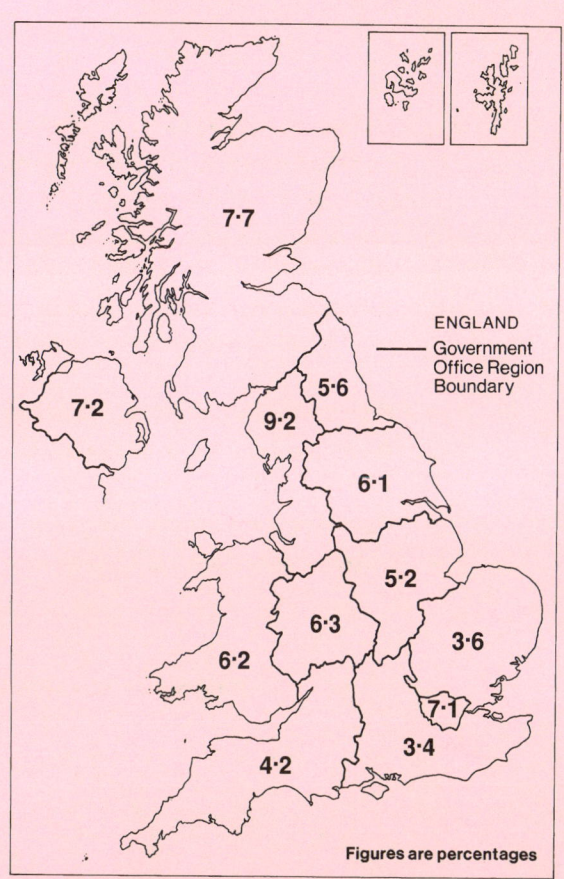

ENGLAND
Government
Office Region
Boundary

7·7

7·2

5·6

9·2

6·1

5·2

6·3

3·6

6·2

7·1

4·2

3·4

Figures are percentages

Reference Map Q5D
(Average gross weekly earnings, April 2000)

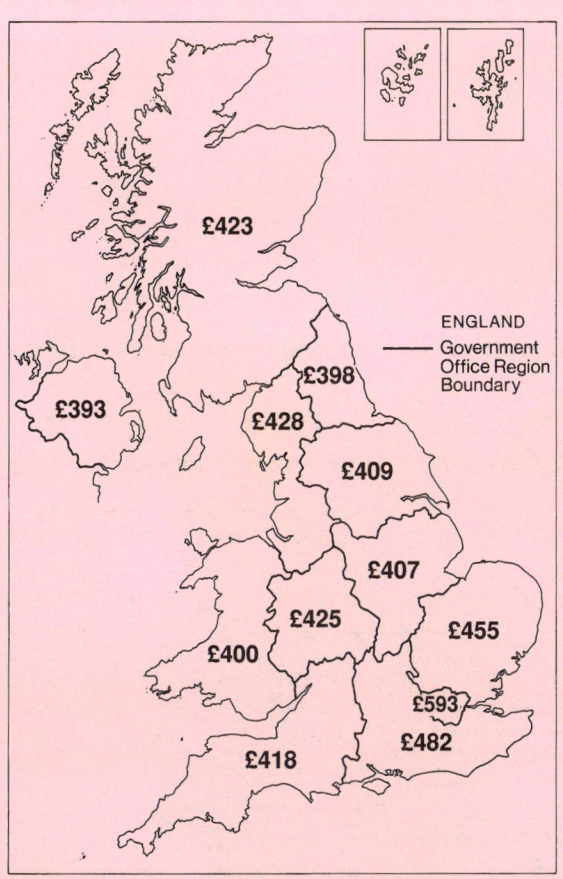

ENGLAND
Government
Office Region
Boundary

£423

£393

£398

£428

£409

£407

£425

£455

£400

£593

£482

£418

Reference Map Q5E
(Cars per 1000 people, 2000)

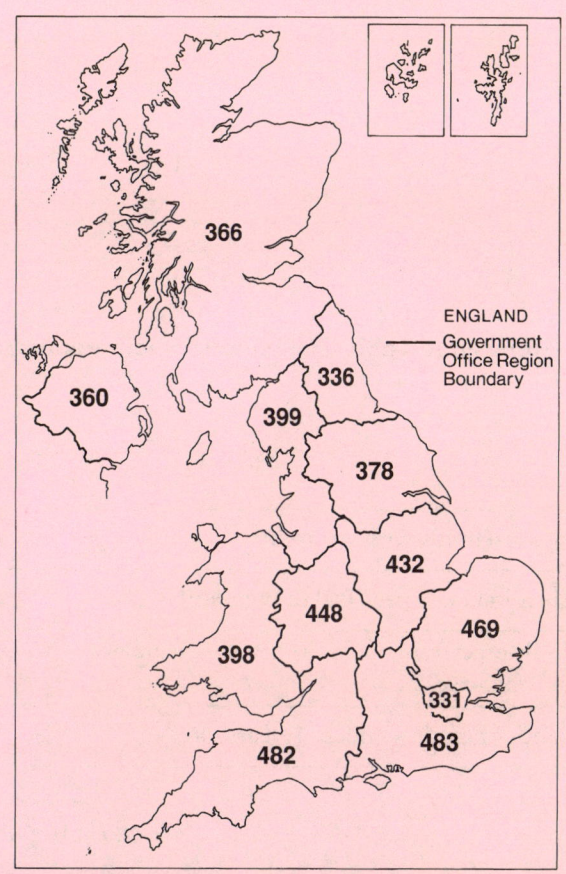

ENGLAND
Government
Office Region
Boundary

366

360

336

399

378

432

448

469

398

331

482

483

Marks

Question 6 (Development and Health)

(a) "*Levels of wealth and economic development are not evenly spread* **within** *individual countries.*"

Study Reference Map Q6 and Reference Table Q6.

  (i) **In what ways** does the information given in the table suggest that the five regions of Brazil are at different levels of development? **5**

  (ii) For Brazil **or** any other **Developing** country which you have studied, **suggest reasons** why such regional variations exist. **5**

  (iii) **Explain** why indicators of development such as those shown in the table may fail to provide an accurate representation of the true quality of life within an area. **3**

(b) For **either** malaria **or** bilharzia (schistosomiasis) **or** cholera,

  (i) **describe** the environmental **and** human factors which put people at risk of contracting the disease,

  (ii) **describe** the methods used to try to control the spread of the disease, and

  (iii) **comment** on how effective these methods have been. **12**

(25)

Reference Map Q6 (Regions of Brazil)

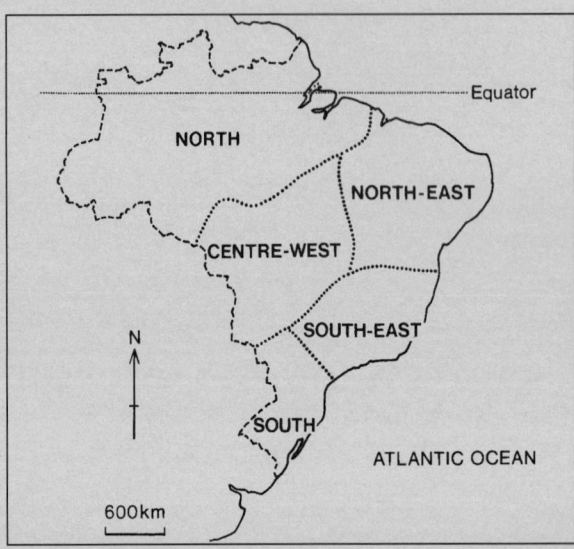

Reference Table Q6 (Selected socio-economic indicators of development for Brazil's regions)

	South-east	South	Centre-west	North-east	North
% households with electricity	98·0	97·0	93·0	78·0	60·0
Hospital beds per 1000 of population	4·2	4·1	4·3	3·1	2·3
% illiteracy rate of children 10–14 years old in urban areas	1·7	1·7	2·1	13·0	5·9
% of population under 18 years old	37·0	38·0	42·0	47·0	48·0

[END OF QUESTION PAPER]

[BLANK PAGE]

X042/301

NATIONAL
QUALIFICATIONS
2004

MONDAY, 17 MAY
9.00 AM – 10.30 AM

GEOGRAPHY
HIGHER
Core

Attempt **all** questions.

The value attached to each question is shown in the margin.

Credit will be given for appropriate models, diagrams, maps and graphs.

Marks may be deducted for bad spelling, bad punctuation and for writing that is difficult to read.

Note The reference maps and diagrams in this paper have been printed in black only: no other colours have been used.

SCOTTISH
QUALIFICATIONS
AUTHORITY

©

Extract No 1350/172

1:50 000 Scale
Landranger Series

Four colours should appear above; if not then please return to the invigilator.
Four colours should appear above; if not then please return to the invigilator.

Scale 1:50 000

2 centimetres to 1 kilometre (one grid square)

Marks

Question 1

"*Tropical latitudes receive more solar energy than Polar latitudes. The atmosphere and oceans help to redistribute this energy to maintain a global energy balance.*"

(*a*) **Explain** fully the ways in which the circulation cells in the atmosphere help to redistribute energy.

4

(*b*) Study Reference Map Q1.

For **either** the Pacific Ocean **or** the Atlantic Ocean, **explain** how the ocean currents operate to maintain the energy balance.

3

Reference Map Q1 (Ocean currents)

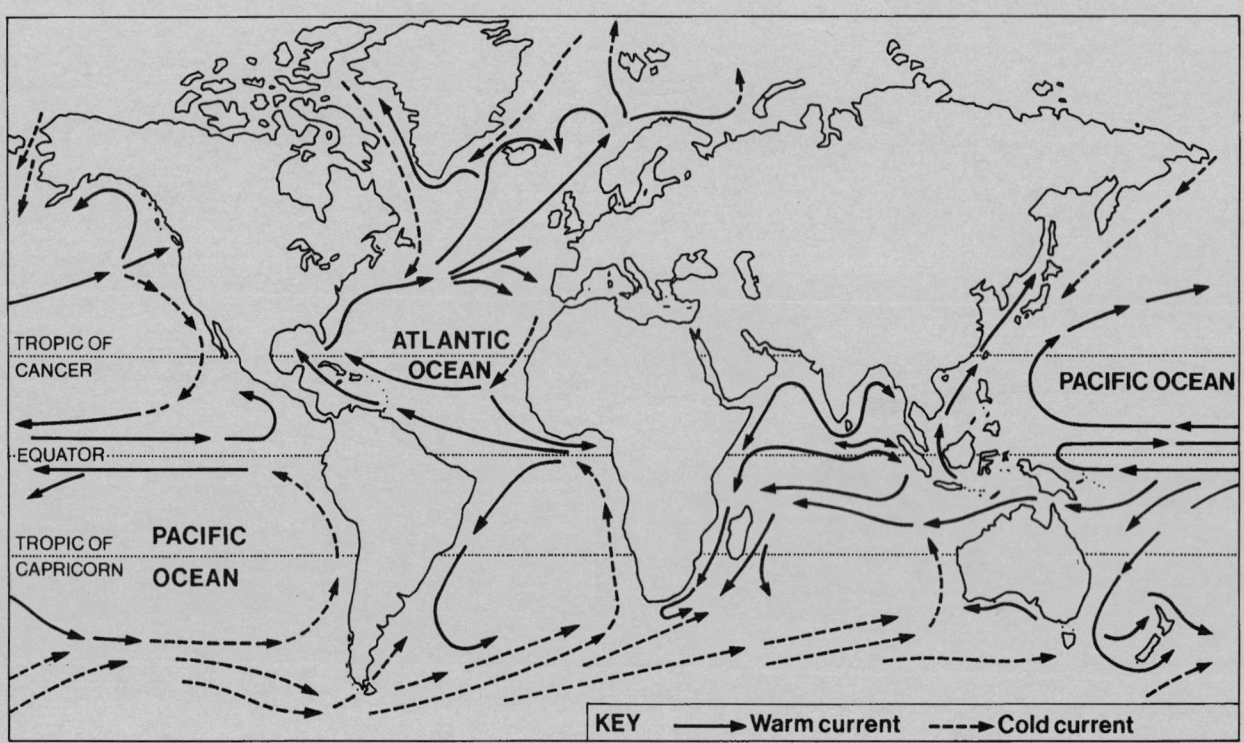

Marks

Question 2

(*a*) With the aid of a diagram, **describe** the global hydrological cycle.　　3

(*b*) Study Reference Diagram Q2.

Explain the changes in **either** erosional processes **or** depositional processes from the upper to the lower part of a drainage basin.　　3

Reference Diagram Q2 (Block diagram of a drainage basin)

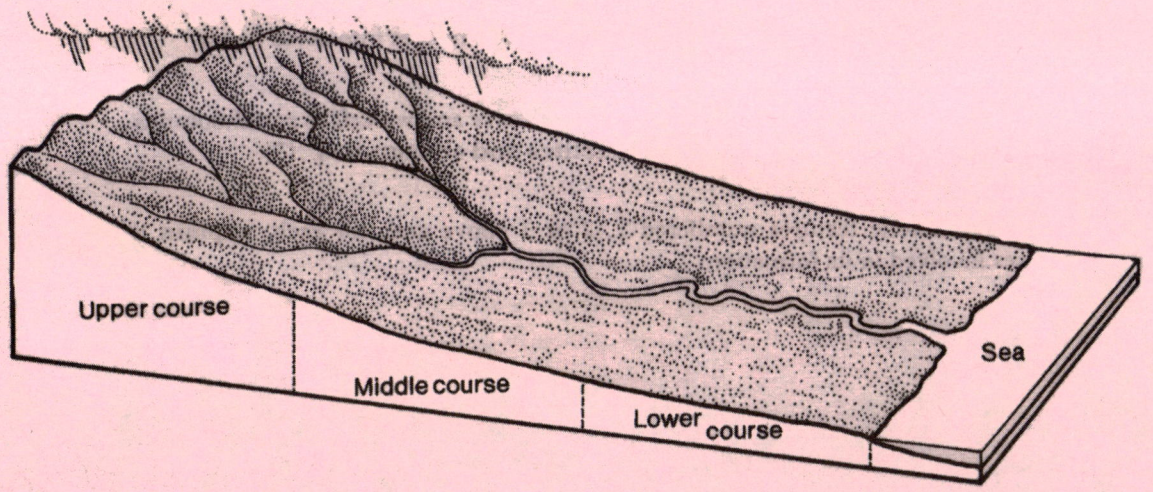

Upper course

Middle course

Lower course

Sea

[Turn over

Question 3

The Yorkshire Dales National Park is an area famous for its Carboniferous Limestone scenery. The following descriptions were taken from the official guide to the Park.

Tourist Attraction	Geographical Description
Malham Cove	80 metre cliffs topped by a fascinating **limestone pavement**. (1)
Gaping Gill Pothole	A magnificent **swallow hole** (2) with water dropping over 100 metres into caves below.
White Scar Caves	Underground cave and waterfalls and thousands of **stalactites and stalagmites**. (3)

Select **two** of the limestone features numbered 1, 2 or 3 above and, for each, **explain** the physical processes involved in its formation.

6

Marks

Question 4

Choose **one** of the following soil types:

* podzol;
* brown forest soil;
* gley.

(*a*) **Draw** a soil profile and annotate it to show the main characteristics of the soil.

(*b*) **Explain** the processes which have created this soil profile.

6

[Turn over

Marks

Question 5

Study Reference Table Q5 which shows some of the major population changes which took place in the UK during the twentieth century.

(*a*) **Suggest reasons** for the changes shown in the table.   **4**

(*b*) **Describe** some of the problems which the Government will have to deal with as a result of an ageing population.   **3**

Reference Table Q5 (Twentieth-century population changes in the UK)

Population Characteristic	Early 1900s	Late 1990s
Population	38 million	59 million
Birth rate	26 per thousand	12 per thousand
Death rate	14 per thousand	11 per thousand
Infant mortality rate	143 per thousand	6 per thousand
Proportion of population over 50 years of age	17%	32%

Marks

Question 6

Study Reference Map Q6.

Choose **one** of the traditional farming systems shown on the map. Referring to a named area where this type of farming is carried out, **assess** the impact which recent changes have had on the people, their way of life and the farming landscape.

6

Reference Map Q6 (Generalised distribution of selected agricultural systems)

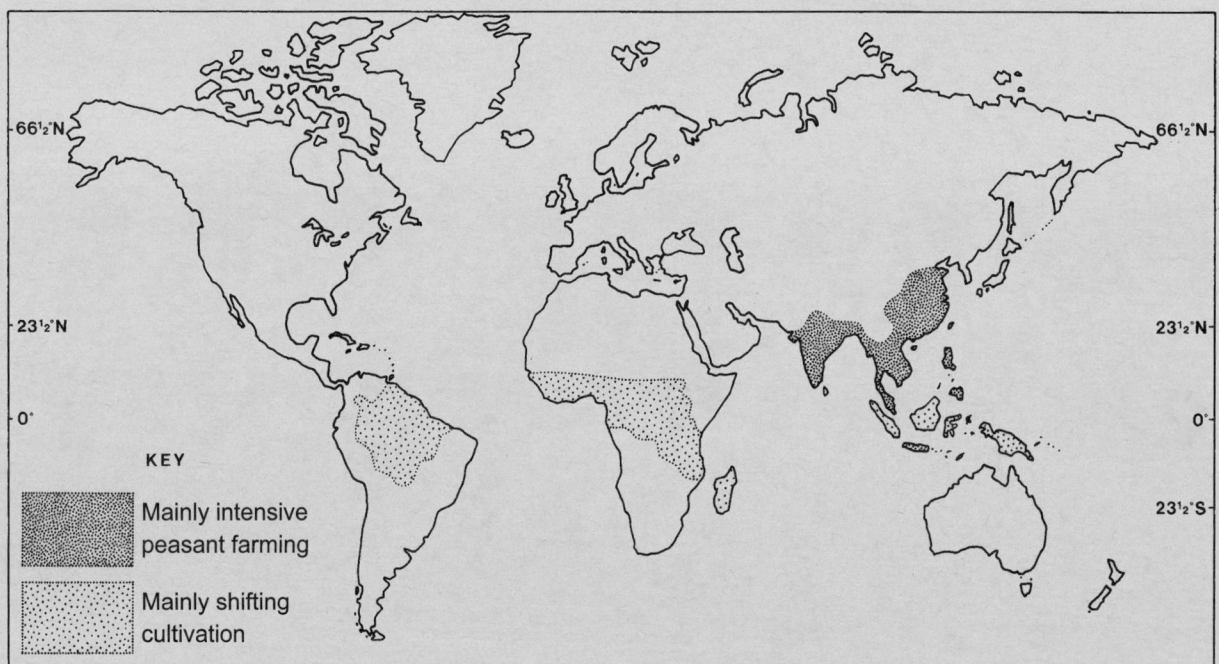

[Turn over

Marks

Question 7

Study OS map extract number 1350/172: Bristol (*separate item*), and Reference Map Q7.

Using map evidence, **describe** and **explain** the physical and human factors that encouraged industry to locate in Area A.

6

Reference Map Q7

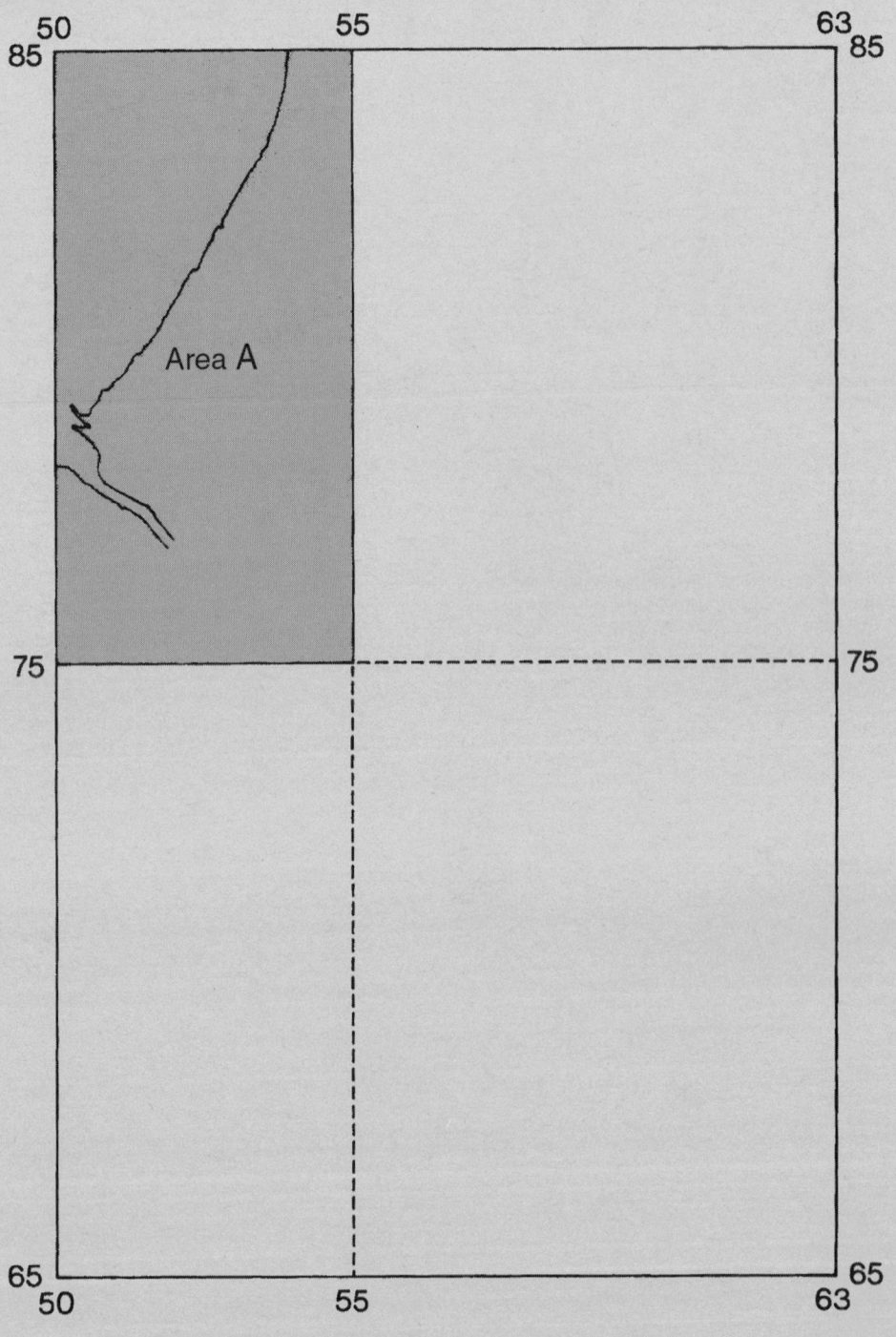

Marks

Question 8

Study OS map extract number 1350/172: Bristol (*separate item*), and Reference Map Q8.

Zones 1 and 2 contain contrasting residential environments.

Describe the residential environments of Zone 1 and Zone 2 and **suggest reasons** for the differences.

6

Reference Map Q8

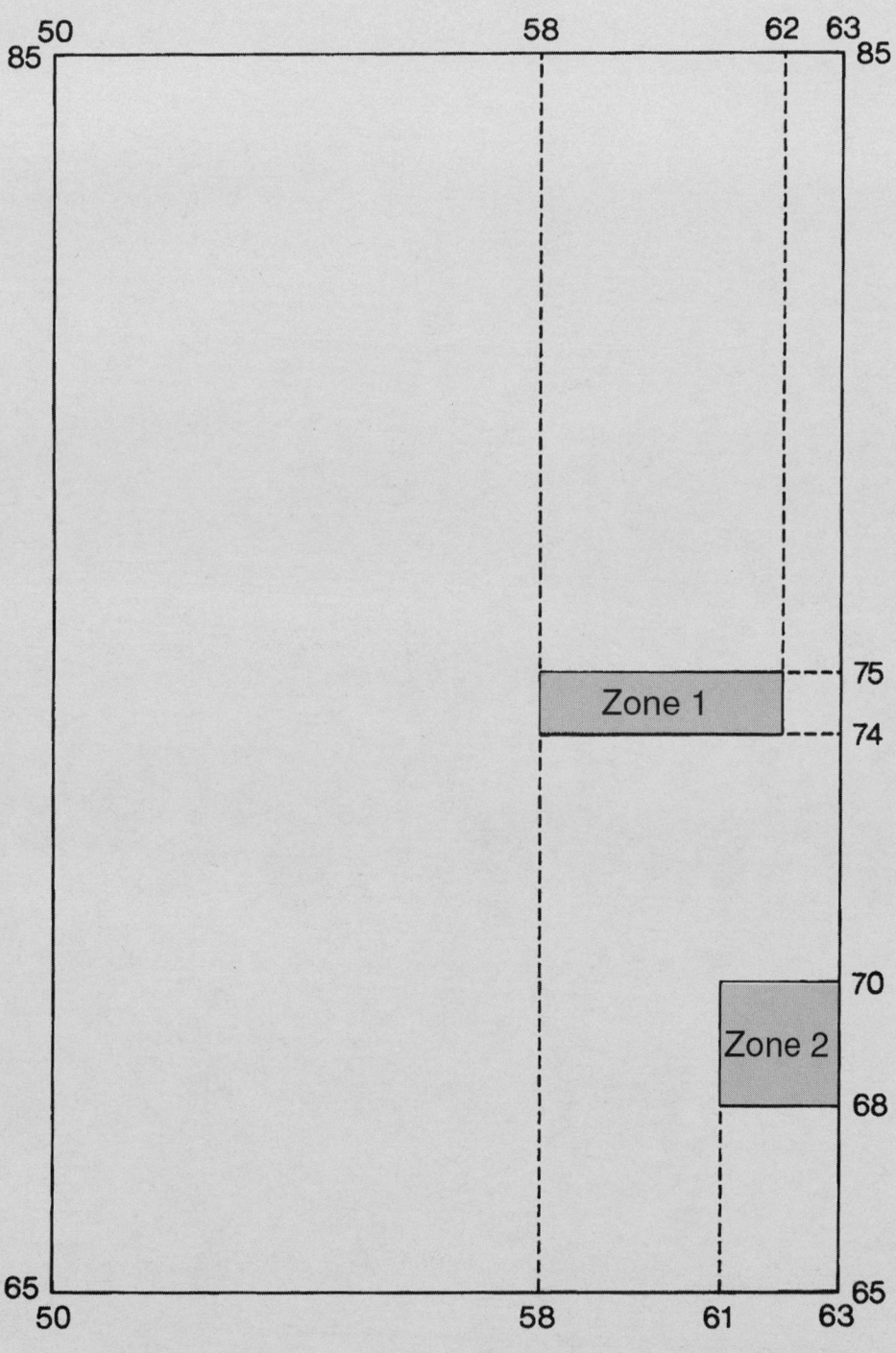

[END OF QUESTION PAPER]

[BLANK PAGE]

X042/303

NATIONAL
QUALIFICATIONS
2004

MONDAY, 17 MAY
10.50 AM – 12.05 PM

GEOGRAPHY
HIGHER
Applications

Two questions should be attempted.

One question from Section 1 (Questions 1, 2, 3) and
one question from Section 2 (Questions 4, 5, 6).

Write the numbers of the **two** questions you have attempted in the marks grid on the back cover of your answer booklet.

The value attached to each question is shown in the margin.

Credit will be given for appropriate maps and diagrams, and for reference to named examples.

Marks may be deducted for bad spelling, bad punctuation and for writing that is difficult to read.

Note The reference maps and diagrams in this paper have been printed in black only: no other colours have been used.

SCOTTISH
QUALIFICATIONS
AUTHORITY

©

Marks

SECTION 1

You must answer ONE question from this Section.

Question 1 (Rural Land Resources)

(a) Study Reference Diagram Q1.

The Cairngorms is an area of outstanding glaciated upland scenery and has been designated as Scotland's second National Park.

Describe and **explain**, with the aid of annotated diagrams, the formation of the glacial features in the Cairngorms or in any glaciated upland area in the UK which you have studied. **10**

(b) **Explain** the ways in which environmental factors limit social and economic opportunities in the Cairngorms **or** in any upland area in the UK which you have studied. **5**

(c) Study Reference Map Q1.

Environmental conflicts may occur in recreational areas in upland landscapes, eg Loch Morlich in the Cairngorms.

With reference to any such recreational area you have studied,

(i) **describe** and **explain** the environmental conflicts, and

(ii) **explain** the ways in which the protection of National Park status might help resolve environmental conflicts. **10**

(25)

Reference Diagram Q1 (The Cairngorms mountain range)

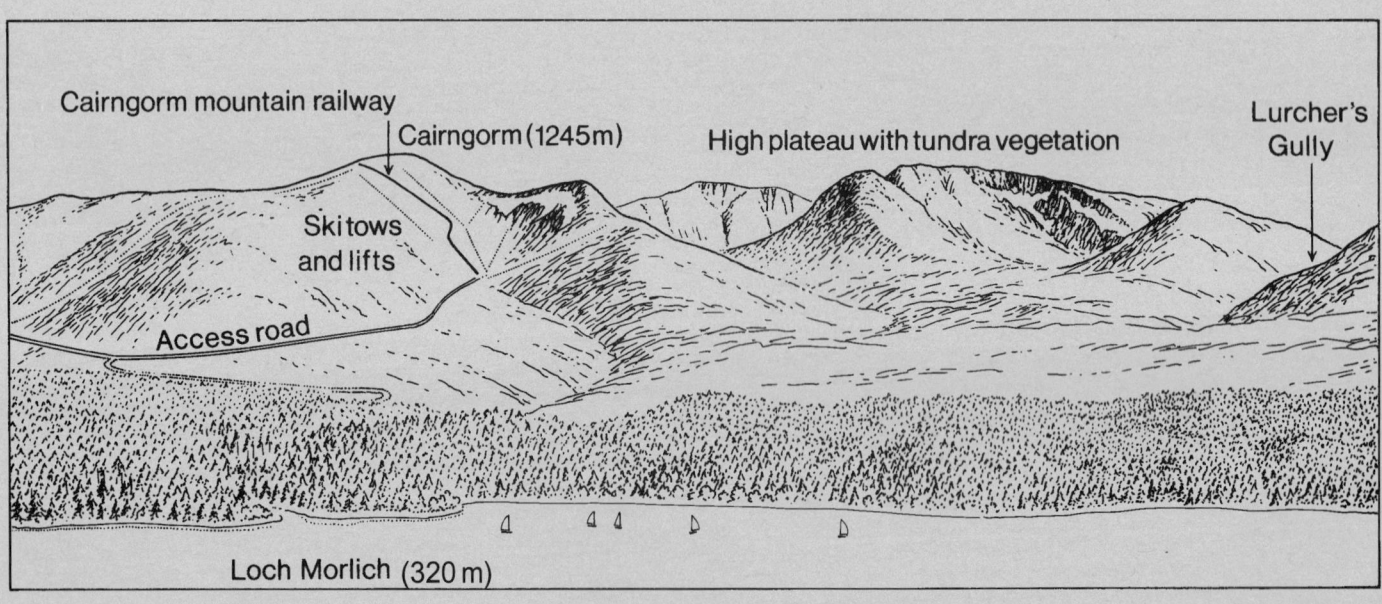

Question 1 – continued

Reference Map Q1 (Tourist map of the Loch Morlich/Cairngorms area)

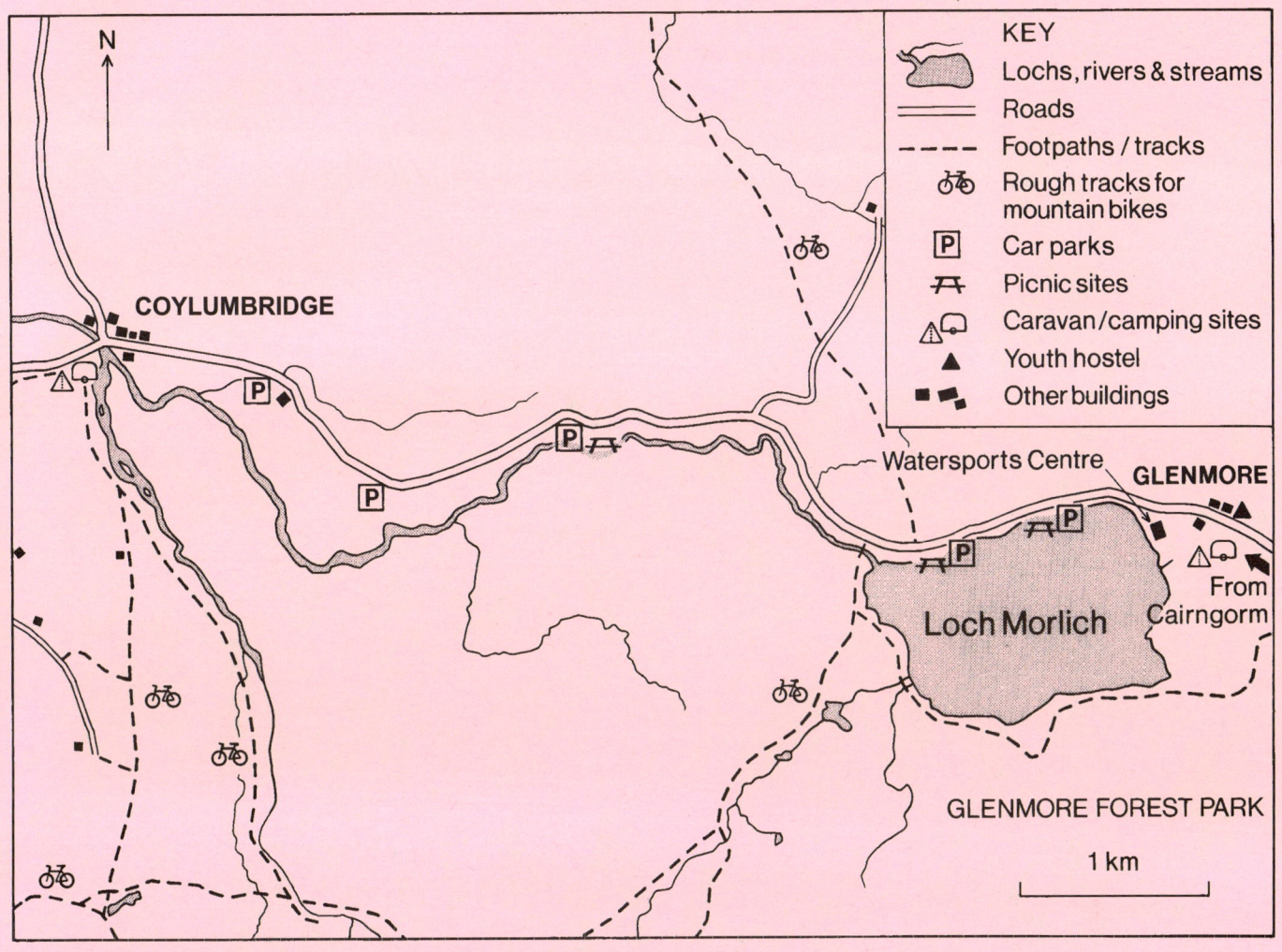

[Turn over

Marks

Question 2 (Rural Land Degradation)

(*a*)   (i)   Study Reference Map Q2 and Reference Diagram Q2. **Explain** how rainfall variability such as that shown on Reference Diagram Q2 may contribute to land degradation in North America.

          3

      (ii)   Outline the ways in which **human** activities have contributed to land degradation in any named area (or areas) of North America.

          5

(*b*) **EITHER**

      (i)   **Describe** the impact of rural land degradation on settled and nomadic people and their environments in Africa North of the Equator.

          8

    OR

      (ii)   **Describe** the impact of rural land degradation on shifting cultivators, ranchers, and the environment in the Amazon Basin.

          8

(*c*) **Describe** the methods used to conserve soil in North America and **evaluate** their effectiveness.

          9

          (25)

Question 2 – continued

Reference Map Q2 (Great Plains states, showing location of South Dakota)

Reference Diagram Q2 (Rainfall variability in South Dakota 1916–1945)

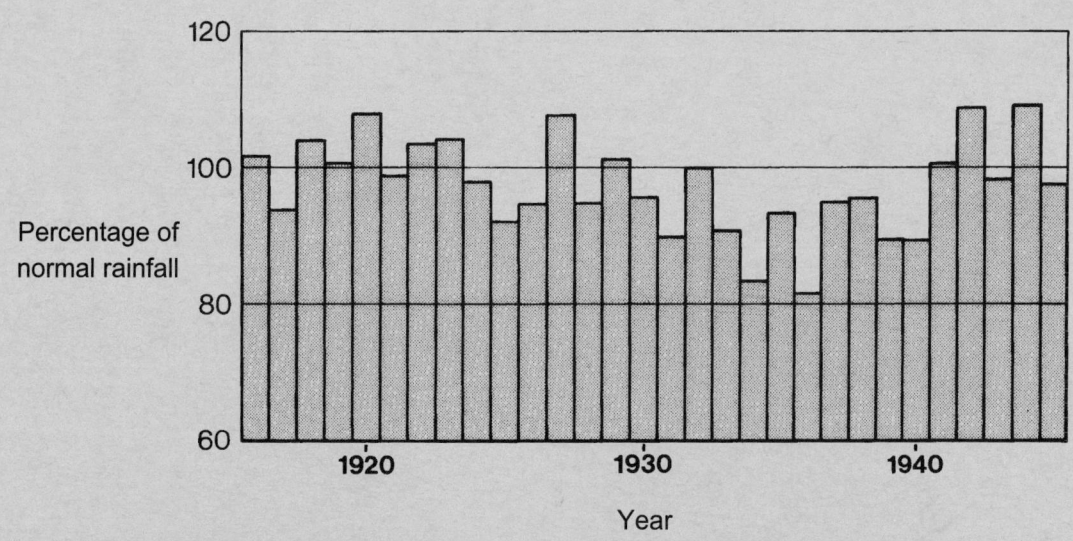

Marks

Question 3 (River Basin Management)

(*a*) Study Reference Maps Q3A and Q3B and Reference Diagrams Q3A and Q3B.

   (i) **Explain** why there is a need for water management in Southern Africa.   **4**

   (ii) **Explain** the ways in which the relief and climate of the area may assist the development of the Lesotho Highlands Water Project.   **4**

(*b*) For the Lesotho Highlands Water Project **or** water control projects in an African or North American river basin you have studied, **describe and account for** the social, economic and environmental benefits **and** adverse consequences.   **12**

(*c*) The Lesotho Highlands Water Project is one of many around the world which cross the political boundaries of two or more countries or states.

Giving examples from Africa **or** North America, **describe** the problems which this can lead to and suggest ways in which they may be overcome.   **5**

                                                        (25)

Reference Map Q3A (Southern Africa)　　　**Reference Map Q3B (Lesotho and neighbouring countries)**

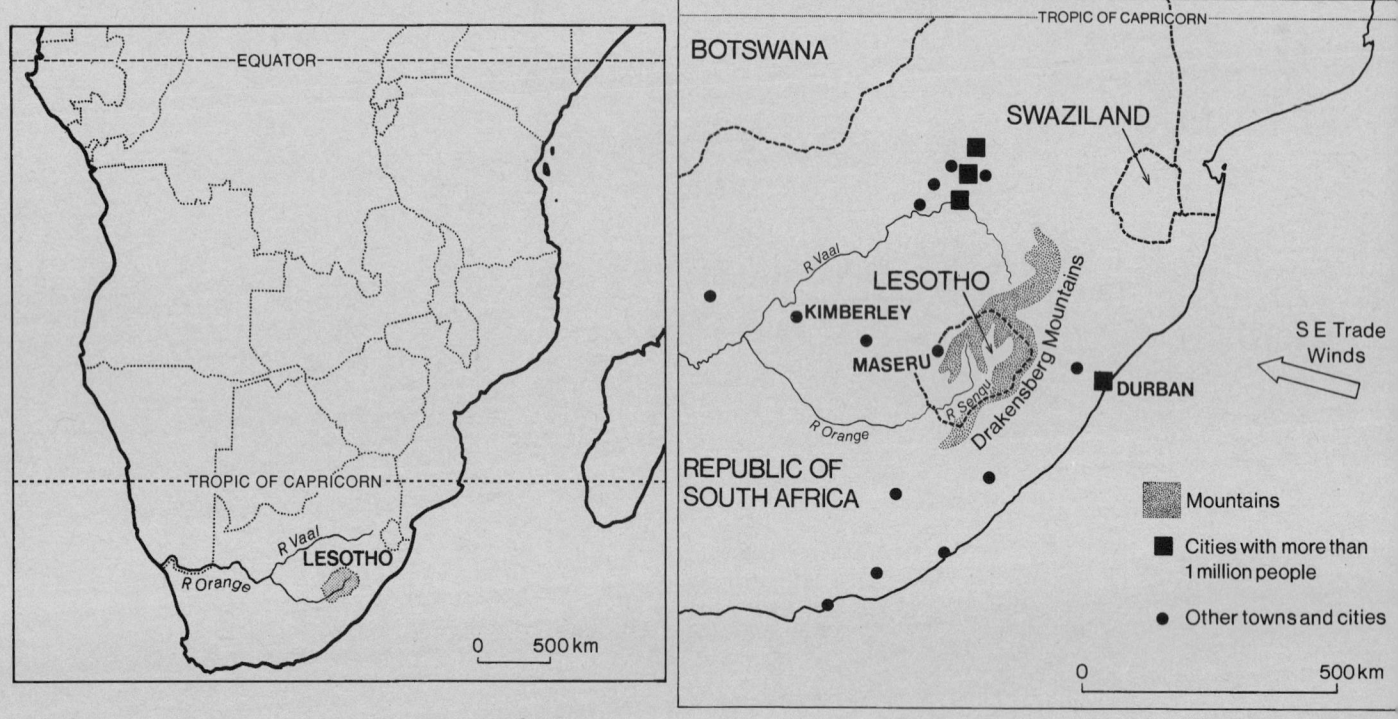

Question 3 – continued

Reference Diagram Q3A (Rainfall graphs for selected stations in Southern Africa)

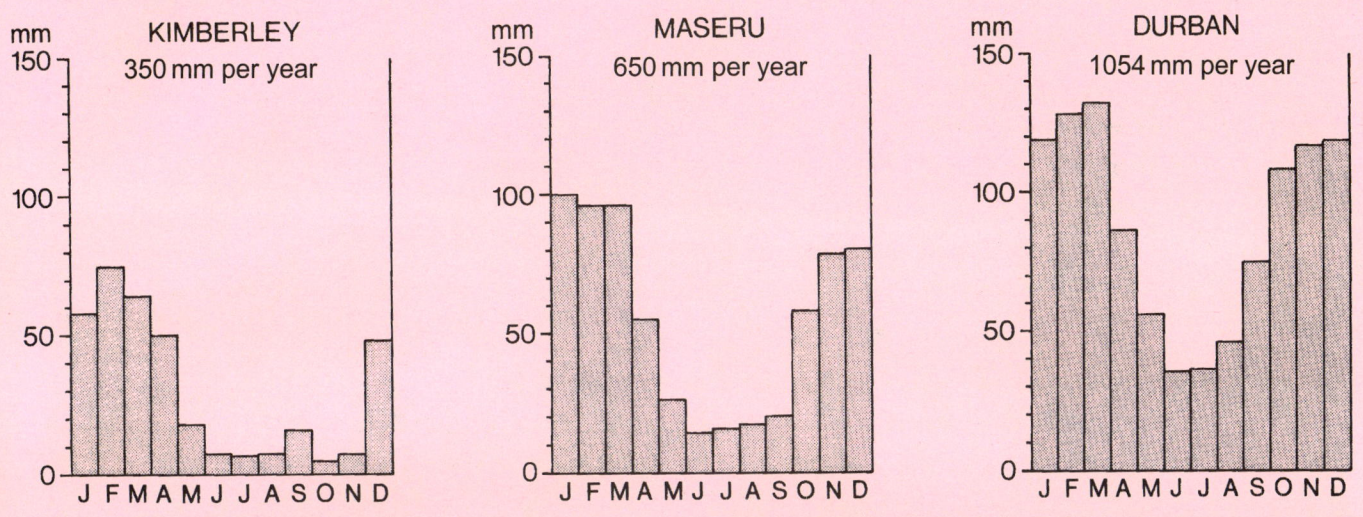

Reference Diagram Q3B (The Lesotho Highlands Water Project)

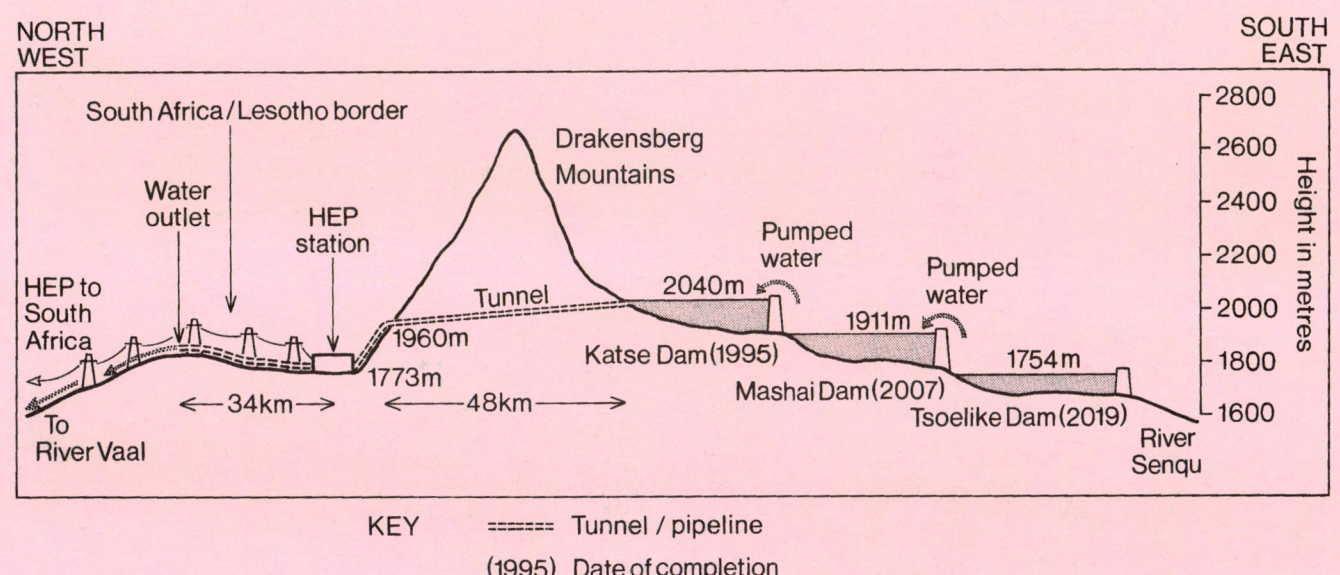

[Turn over

SECTION 2

You must answer ONE question from this Section.

Question 4 (Urban Change and its Management)

(*a*) In the second half of the 20th century, Central Business Districts of cities in countries in the **Developed World** (EMDCs) have undergone major changes in:

- shopping
- transport
- entertainment/leisure.

Choose **two** of these and, referring to a city you have studied, **describe** and **explain** these changes.

8

(*b*) Referring to a city you have studied in the **Developed World,**

(i) **describe** and **explain** the conflicts in land-use on the edge of the city, and

(ii) **outline** the strategies adopted by the city authorities to resolve these conflicts.

8

(*c*) Study Reference Diagram Q4.

Many cities in the **Developing World** (ELDCs) continue to grow rapidly.

For Mumbai (Bombay) or any city you have studied in the **Developing World,**

(i) **account** for its rapid growth, and

(ii) **identify** the social, economic and environmental problems that have resulted from its rapid growth.

9

(25)

Reference Diagram Q4 (Population of Mumbai: formerly Bombay)

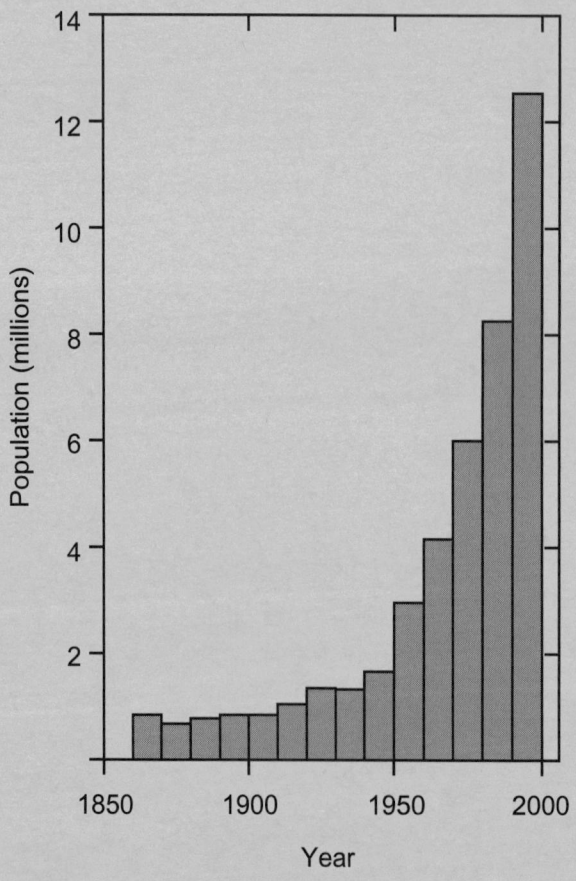

Marks

Question 5 (European Regional Inequalities)

Study Reference Diagram Q5.

(a) Referring to specific examples, **describe** the regional inequalities

    (i) **between** the member states shown, and

    (ii) **within** the member states shown.   **5**

(b) For any one **named** EU country you have studied,

    (i) **describe** the physical and human factors that have contributed to the regional inequalities, and   **7**

    (ii) **outline** the social and economic problems that can result from these inequalities.   **5**

(c) **Describe** the strategies used by the European Union and national governments to reduce these inequalities and **comment** on their effectiveness.   **8**

    (25)

Reference Diagram Q5

Country	Wealthiest region of country		Poorest region of country		National average
	Regional disparities in purchasing power* between selected Member States (1997) (EU average = 100)				
	Name	Purchasing power	Name	Purchasing power	
Belgium	Brussels	170	Hainaut	80	112
Germany	Hamburg	195	Dessau	52	110
Spain	Balearics	100	Extremadura	50	79
France	Ile-de France (includes Paris)	160	Languedoc-Roussillon	80	101
Ireland	Southern and Eastern (includes Dublin)	101	Border, Midland and Western	68	96
Italy	Lombardia	130	Calabria	55	103
Portugal	Lisbon and Vale do Tejo	94	Alentejo	47	71
United Kingdom	Inner London	220	Cornwall and Isles of Scilly	70	99

*Purchasing power is based on cost of living in a region or country.

Marks

Question 6 (Development and Health)

(*a*)   (i)   There are considerable differences in levels of development **between** countries in the **Developing World** (ELDCs).

Referring to countries in the **Developing World** which you have studied, **suggest reasons** why such wide variations in development exist **between** countries.   **6**

(ii)   Identify one social **and** one economic indicator of development and, for each, **explain** how it might illustrate a country's level of development.   **4**

(*b*)   Study Reference Diagram Q6 which highlights the main aspects of Primary Health Care (PHC).

Give examples of Primary Health Care strategies and suggest why such approaches to improving health standards are suited to **Developing Countries**.   **6**

(*c*)   Bilharzia (Schistosomiasis), Cholera and Malaria remain major health problems in many parts of the **Developing World**.

For **one** of these diseases, **describe** and **evaluate** the measures used in attempting to control it.   **9**

   (25)

Reference Diagram Q6 (Elements of Primary Health Care)

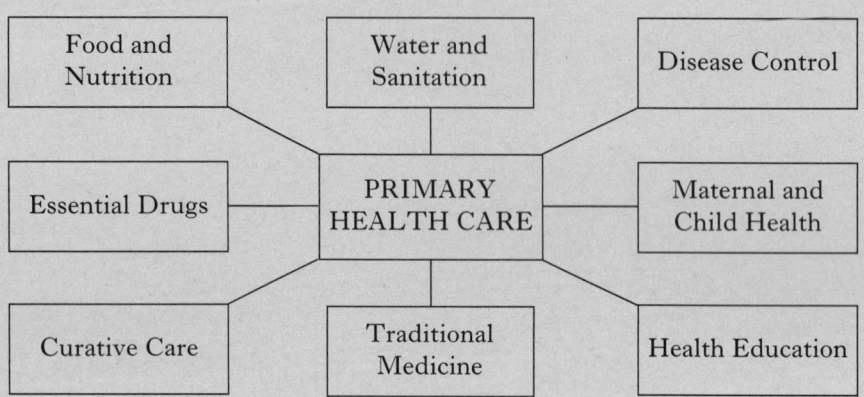

[END OF QUESTION PAPER]

Dear Student

In 2005 the format of the Higher Geography exams was changed. The following Specimen Question Paper and the actual 2005 exam will give you good practice in this new format. However, the previous years' exams in this book will provide just as good revision and exam-practice features.

Here is some information about the new exam format:

- There are two Question Papers, each marked out of 50.

- The time allocation for Paper 1 is 1 hour 30 minutes.

- The time allocation for Paper 2 is 1 hour 15 minutes.

Paper 1 – Physical and Human Environments (Total mark available: 50)
- This paper will consist of three sections, A, B and C.

- The questions will ask you to write short answers that display your knowledge and understanding of the Units Physical Environment and Human Environment.

- At least one question will be based on an Ordnance Survey (OS) map.

- Section A will include four compulsory questions. Two of these will be on Physical Environments topics; two will be on Human Environment topics.

- Section B will include two questions on the Physical Environment topics not examined in Section A. You will have to answer one of these questions.

- Section C will include two questions on the Human Environment topics not examined in Section A. You will have to answer one of these questions.

Paper 2 – Environmental Interactions (Total mark available: 50)
- This paper will consist of two sections, A and B.

- Section A will contain three questions from the Group 1 Environmental Interactions, each marked out of 25.

- Section B will contain three questions from the Group 2 Environmental Interactions, each marked out of 25.

- You will have to answer one question from each section.

Please visit *www.sqa.org.uk* for further details.

[BLANK PAGE]

C208/SQP224

Geography

Higher

Physical and

Human Environments

Specimen Question Paper

for use in and after 2005

Time: 1 hour 30 mins

NATIONAL QUALIFICATIONS

Six questions should be attempted, namely, all **four** questions in Section A, **one** question from Section B and **one** question from Section C.

The value attached to each question is shown in the margin.

Credit will be given for appropriate models, diagrams, maps and graphs.

Marks may be deducted for bad spelling, bad punctuation and for writing that is difficult to read.

Note The reference maps and diagrams in this paper have been printed in black only: no other colours have been used.

NB The Ordnance Survey Map used in this Specimen Question Paper is the map from the Higher examination in 2002.

SCOTTISH
QUALIFICATIONS
AUTHORITY

1:50 000 Scale
Landranger Series

Four colours should a
Four colours should a

Diagrammatic only

1 kilometre = 0·6214 mile

Extract No 1269/171

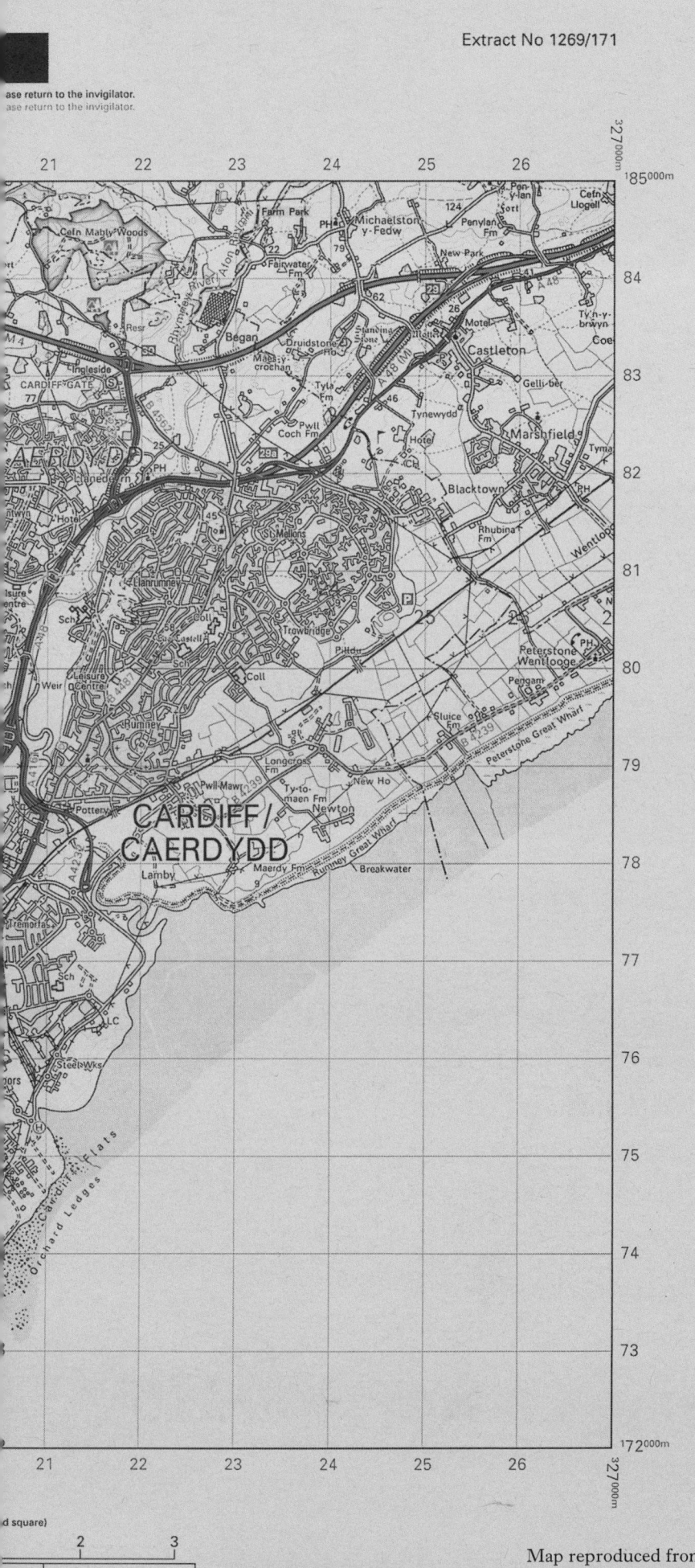

Marks

SECTION A: Answer ALL questions in this section

Question 1: Atmosphere

(*a*)   Study Reference Maps Q1A.

Describe the origin, nature and weather characteristics of the Tropical   **4**
Maritime and Tropical Continental air masses.

(*b*)   Study Reference Maps Q1A and Q1B.

Using the maps and graphs, **describe** and **explain** the pattern of annual   **5**
rainfall in both the north and south of Nigeria.

**Reference Maps Q1A (Location of selected air masses and the ITCZ in
January and July)**

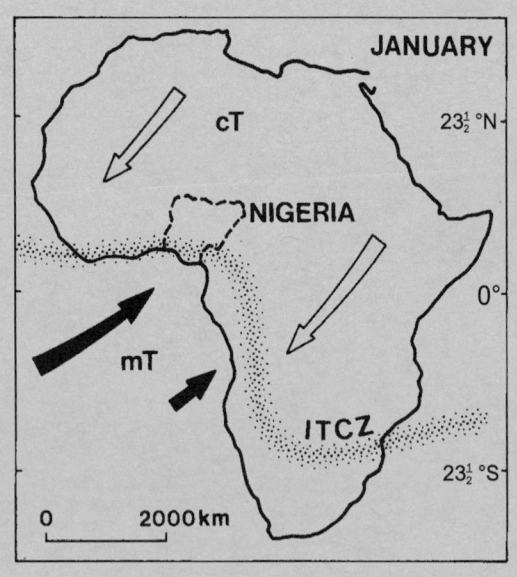 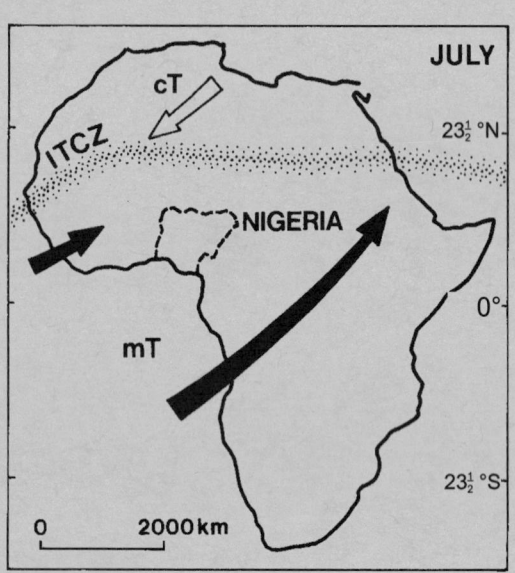

	KEY	
mT	**Tropical Maritime**	
cT	**Tropical Continental**	
ITCZ	**Inter Tropical Convergence Zone**	

Page two

Question 1—continued

Reference Map Q1B (Length, in days, of the rainy season in Nigeria, and selected rainfall graphs)

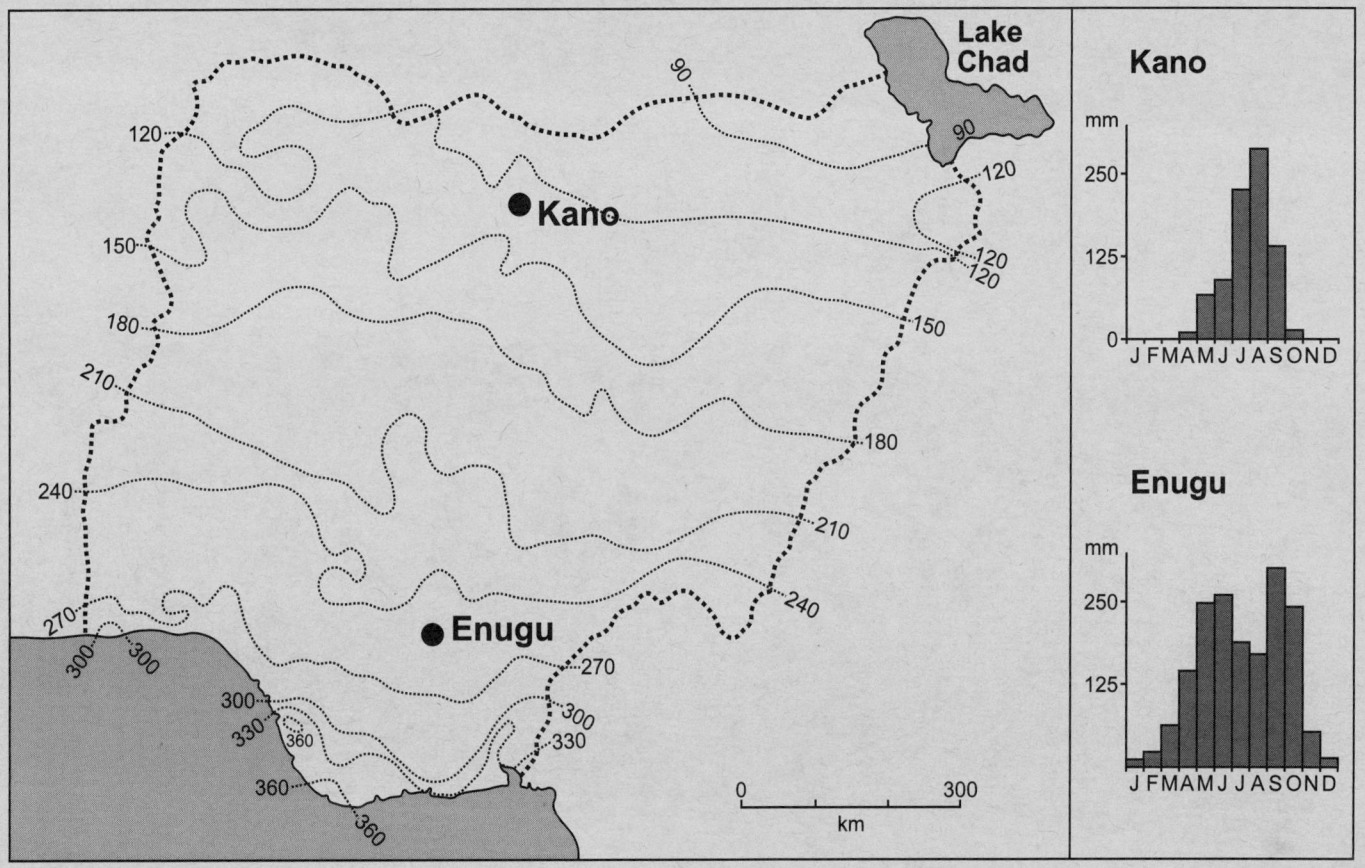

Marks

Question 2: Biosphere

(*a*) **Explain** what is meant by the term "climax vegetation". 2

(*b*) Study Reference Diagram Q2.

Describe and **explain** the plant succession in a sand dune habitat such as that shown in Reference Diagram Q2. You should refer to specific plants. 7

Reference Diagram Q2 (Transect across sand dune coastline)

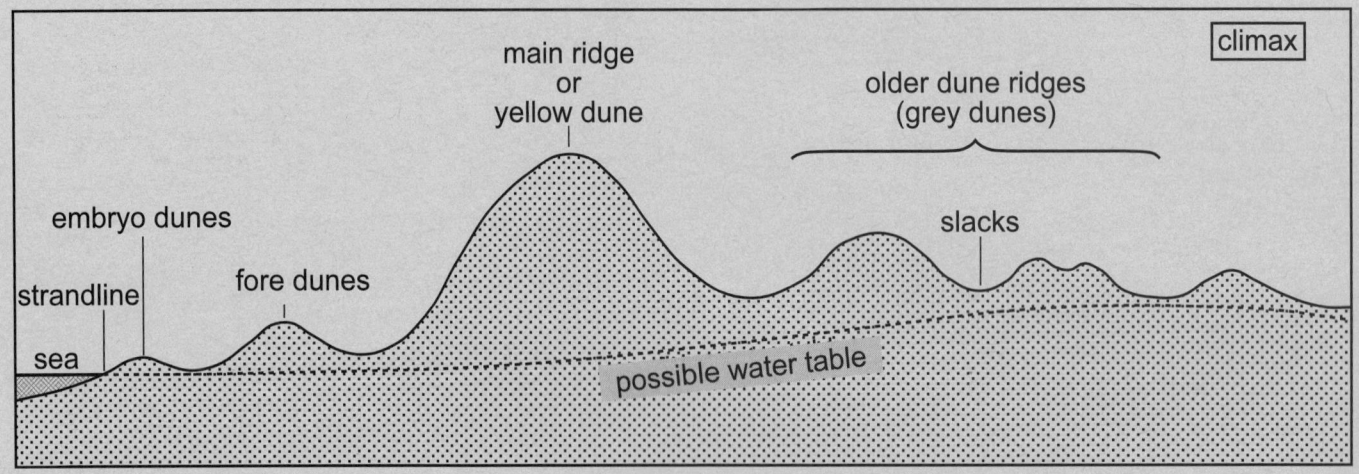

Page four

Marks

Question 3: Rural Geography

Study Reference Table Q3.

(a) Choose **one** of the farming systems from the Table and, referring to a named location, **describe** the system in detail.

5

(b) All of the farming systems have been subjected to change. For any **one** farming system, **describe** the impact of the change on people and the environment.

4

Reference Table Q3 (Farming systems and examples of changes)

Farming System	Example of Change
Shifting Cultivation	Deforestation
Intensive Peasant Farming	Green Revolution
Extensive Commercial Farming	Enlargement of farms and fields

Marks

Question 4: Industrial Geography

Study OS map extract number 1269/171: Cardiff (*separate item*), and Reference Map Q4.

(*a*) Using map evidence, **describe** and **explain** the physical and human factors which encouraged industry to locate in Area A.   **6**

(*b*) **Describe** the likely environmental impact on the surrounding area of the industrial developments in Area A.   **3**

Reference Map Q4

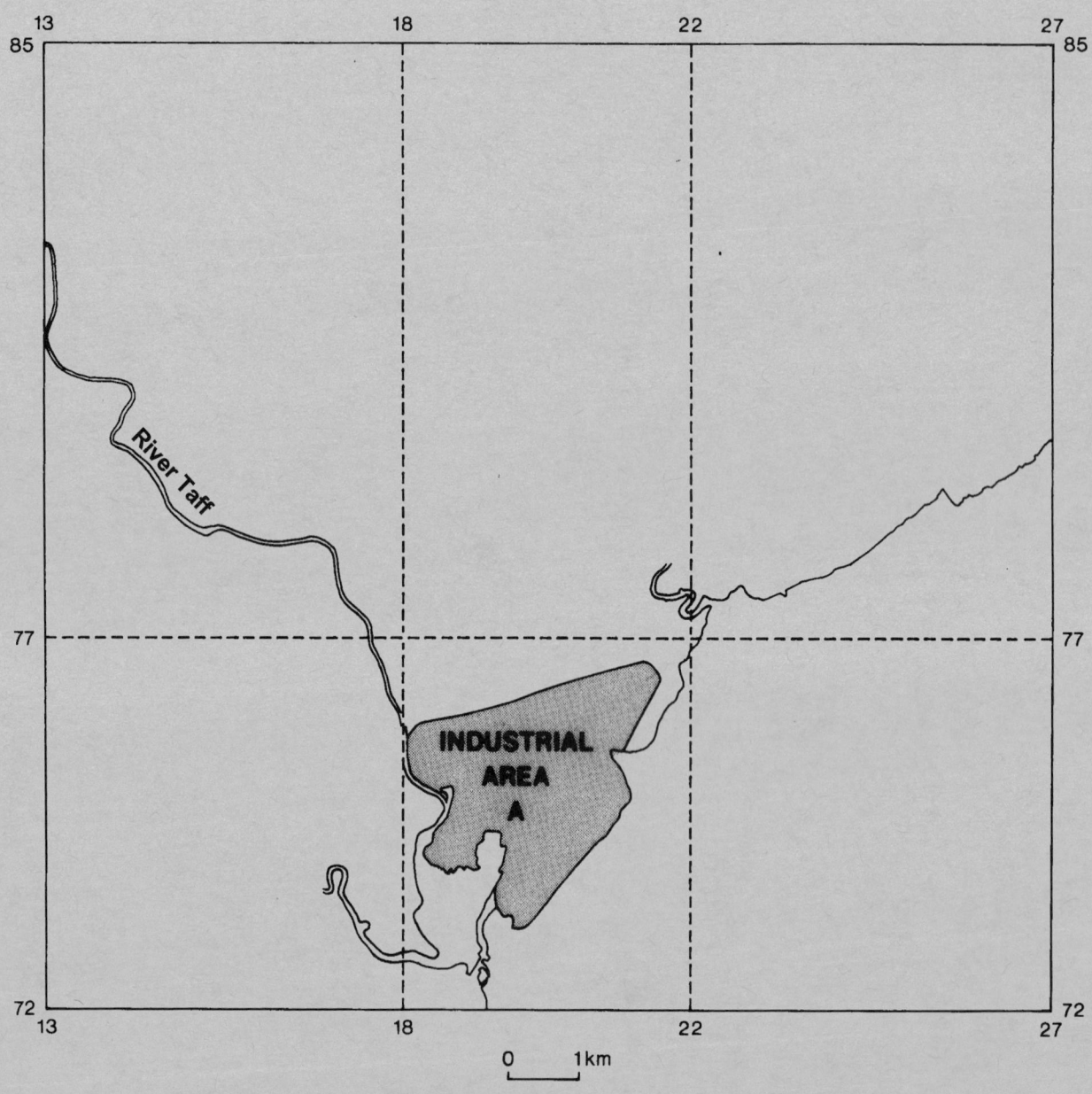

Marks

**SECTION B: Answer ONE question from this section,
ie either Question 5 or Question 6.**

Question 5: Hydrosphere

Study OS map extract number 1269/171: Cardiff (*separate item*).

(*a*)   Using appropriate grid references, **describe** the physical characteristics of the Afon Rhymney (Rhymni) and its valley from 233850 to 223775.   4

(*b*)   Using a diagram or diagrams, **explain** the formation of **one** of the river features you have described in part (*a*).   3

Marks

Question 6: Lithosphere

Study Reference Diagram Q6.

Select **two** features from the following list and **explain** the processes involved in their formation:

  (i)   limestone pavement;

 (ii)   gorge;

(iii)   stalactites and stalagmites.

7

Reference Diagram Q6 (Carboniferous Limestone landscape)

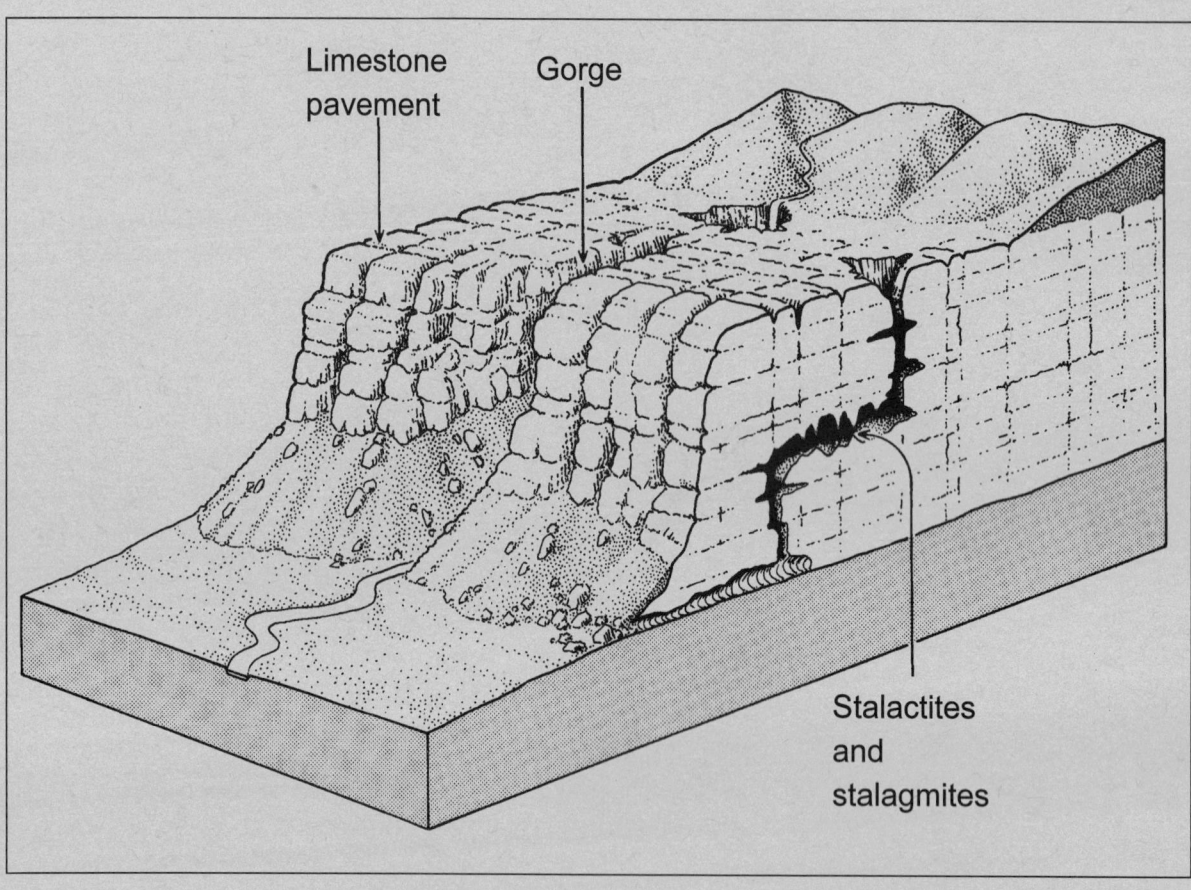

Marks

**SECTION C: Answer ONE question from this section,
ie either Question 7 or Question 8.**

Question 7: Population Geography

Study Reference Diagram Q7.

India has a population structure which is typical of that of many Economically Less Developed Countries.

Describe and **account for** the population structure shown.

7

Reference Diagram Q7 (India: population pyramid 1991)

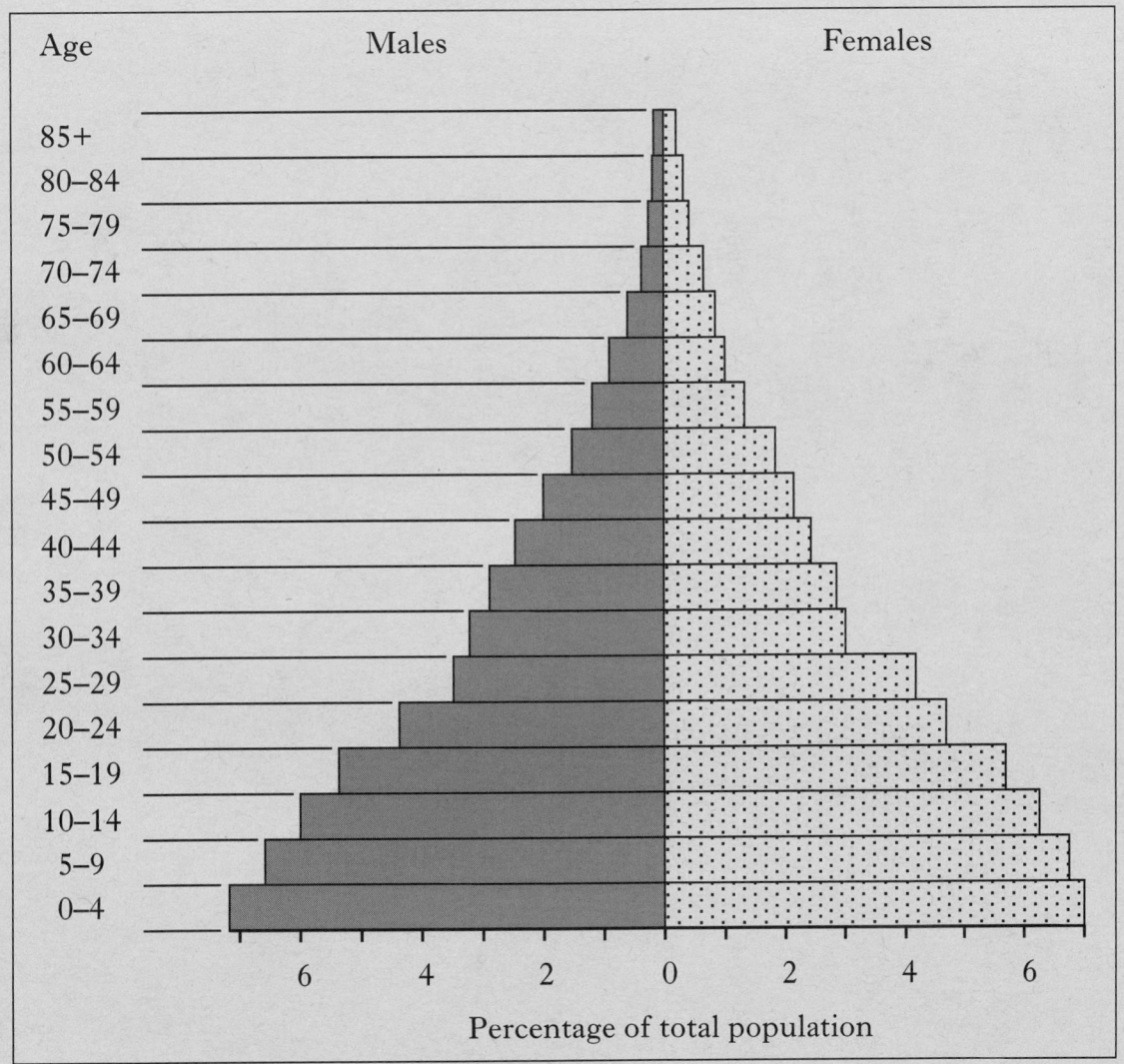

Page nine

Marks

Question 8: Urban Geography

(*a*) For any city you have studied, **show** how its location and site encouraged its growth.

3

(*b*) Study Reference Map Q8 which shows land use zones in Liverpool. Choose **one** of the land use zones A or B identified in the key.

Referring to Liverpool, or any other city you have studied, **describe** and **explain** the changes which have taken place in recent years **in your chosen zone**.

4

Reference Map Q8 (Land use zones in Liverpool)

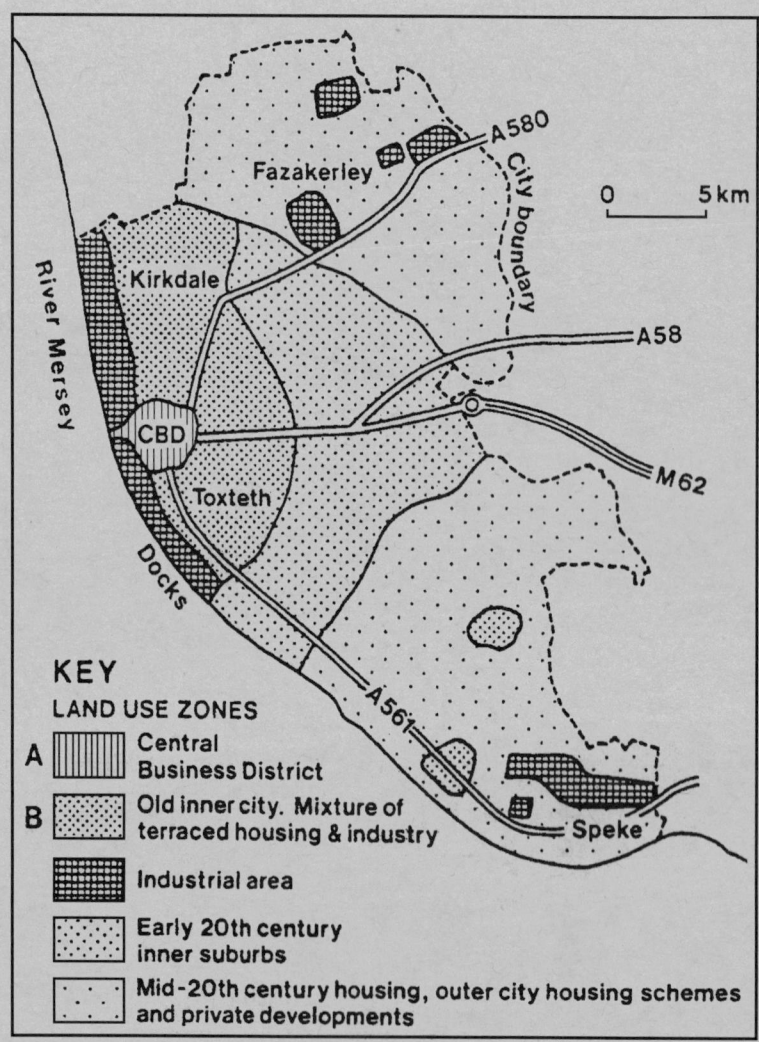

[END OF SPECIMEN QUESTION PAPER]

C208/SQP224

Geography	Time: 1 hour 15mins	NATIONAL
Higher		QUALIFICATIONS

Environmental Interactions

Specimen Question Paper
for use in and after 2005

Two questions should be attempted, namely:
One question from Section 1, (Note: only one question is provided in this Specimen paper).
One question from Section 2, (Questions 2,3,4).

The value attached to each question is shown in the margin.

Credit will be given for appropriate models, diagrams, maps and graphs.

Marks may be deducted for bad spelling, bad punctuation and for writing that is difficult to read.

Questions should be answered in sentences.

Note The reference maps and diagrams in this paper have been printed in black only: no other colours have been used.

SCOTTISH
QUALIFICATIONS
AUTHORITY

©

SECTION 1   *Marks*

Answer Question 1.

Question 1: Rural Land Resources

(*a*)   Study Reference Diagram Q1.

With the aid of annotated diagrams, **describe** the main features of the physical landscape and **explain** the processes involved in their formation.   **10**

(*b*)   For a **named** coastal area you have studied, **explain** the economic and social opportunities provided by the landscape.   **5**

(*c*)   "*Coastal areas within the UK are areas in which environmental conflicts can occur.*"

For any **named** coastal area you have studied,

  (i)   give examples of environmental conflicts which have arisen, and

  (ii) **describe** some of the measures taken to resolve these conflicts and **comment** on their effectiveness.   **10**

                                                                                    (25)

Reference Diagram Q1 (Sketch of selected coastal features)

SEA

SECTION 2

Marks

You must answer ONE question from this Section.

Question 2 (Urban Change and its Management)

(*a*) Study Reference Diagram Q4, which gives information about Lagos, a typical city in the Developing World.

    (i) **Describe** the changes in population growth, built-up area and population density in Lagos between 1890 and 1990.

    (ii) **Comment** on the relationship between these indicators.

5

(*b*) For Lagos, **or** a named city you have studied in the **Developing** World:

    (i) **describe** the economic, social and environmental problems which have resulted from the growth of the city, and

    (ii) **describe** strategies employed by the city authorities to tackle these problems.

10

(*c*) *"There are many problems on the edges of cities in the **Developed** World, eg housing schemes in need of improvement and growing urban pressures on the rural-urban fringe."*

For a **named** city you have studied in the **Developed** World:

    (i) **outline** the factors which contribute to the problems found on the edges of cities, and

    (ii) **describe** efforts made to solve these problems and **assess** how effective they have been.

10

(25)

Reference Diagram Q4 (Lagos—population, area and density 1890–1990)

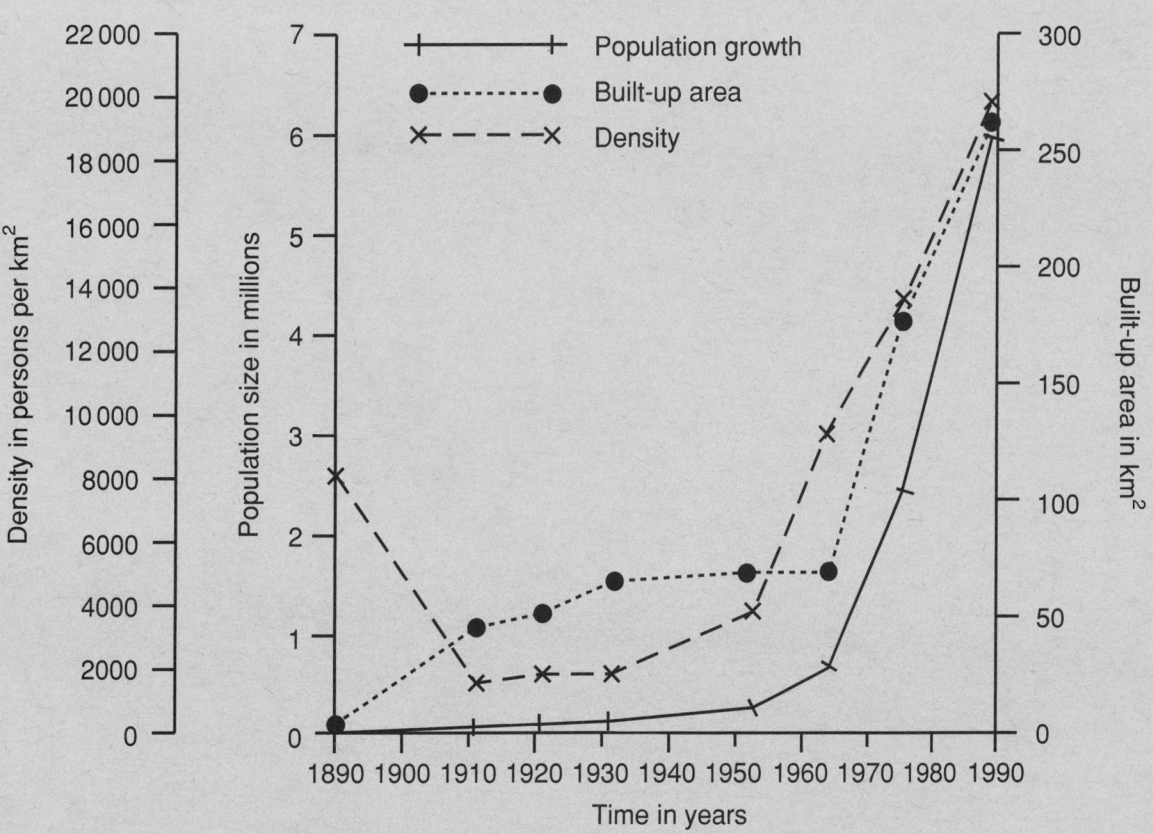

Marks

Question 3 (European Regional Inequalities)

(a) Study Reference Map Q5A.

Suggest both physical **and** human reasons for the prosperity enjoyed within the area now known as "Europe's Hot Banana".

6

(b) Study Reference Map Q5B and Reference Table Q5.

Italy is often described as having a "North-South Divide". With reference to specific provinces and data provided in the table, **comment** on the accuracy of this statement.

5

(c) For Italy or any other country of the European Union which has marked differences in economic development between regions, **describe** the physical **and** human factors which have contributed to such regional differences.

7

(d) For the country chosen in (c), **describe** the strategies used by both the national government and the European Union to overcome these problems, and **comment** on their effectiveness.

7

(25)

Reference Map Q5A (Europe's Hot Banana)

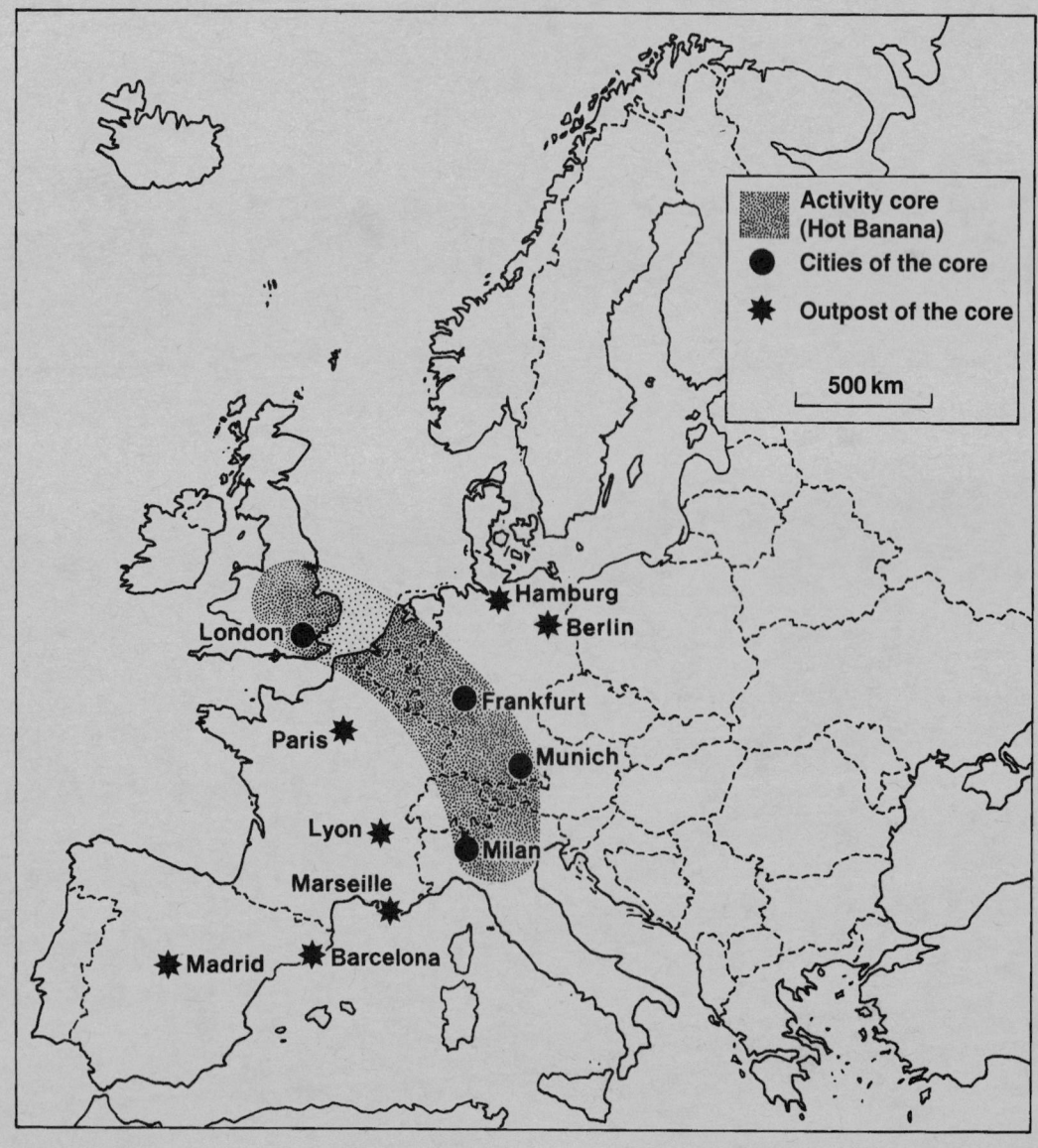

Question 3 – continued

Reference Map Q5B (Italy—GDP per capita, by provinces)

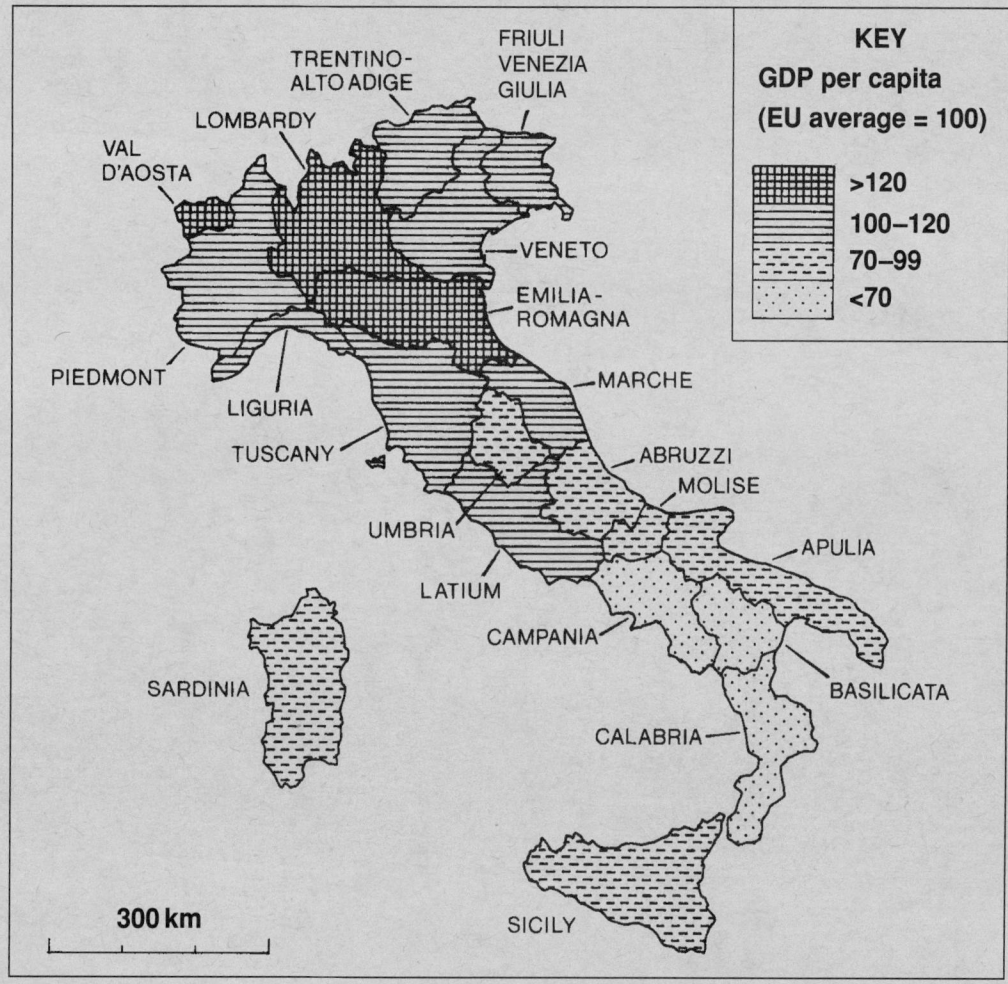

Reference Table Q5 (Italy—Percentage of total employment in selected sectors in North and South)

Employment sector	Percentage in north	Percentage in south
Industry	78	22
Commerce	65	35
Services	59	41

Question 4 (Development and Health)   *Marks*

(*a*) Study Reference Map Q6, which shows the Physical Quality of Life Index (PQLI) for countries of the world.

    (i) For **either** the PQLI **or** any similar composite measure of development, state **three** indicators which might be used in its calculation, and **comment** on the usefulness of the indicators.   **6**

    (ii) Referring to countries in the **Developing World** which you have studied, **suggest reasons** for variations **between** countries in their quality of life.   **5**

    (iii) **Explain** why indicators of development may not accurately reflect the quality of life **within** a country.   **4**

(*b*) For malaria, **or** bilharzia **or** cholera:

    (i) **describe** the methods used to try to control the spread of the disease.   **10**

    (ii) **comment** on how effective these methods of control have been.

                                             (25)

Reference Map Q6 (Physical Quality of Life Index)

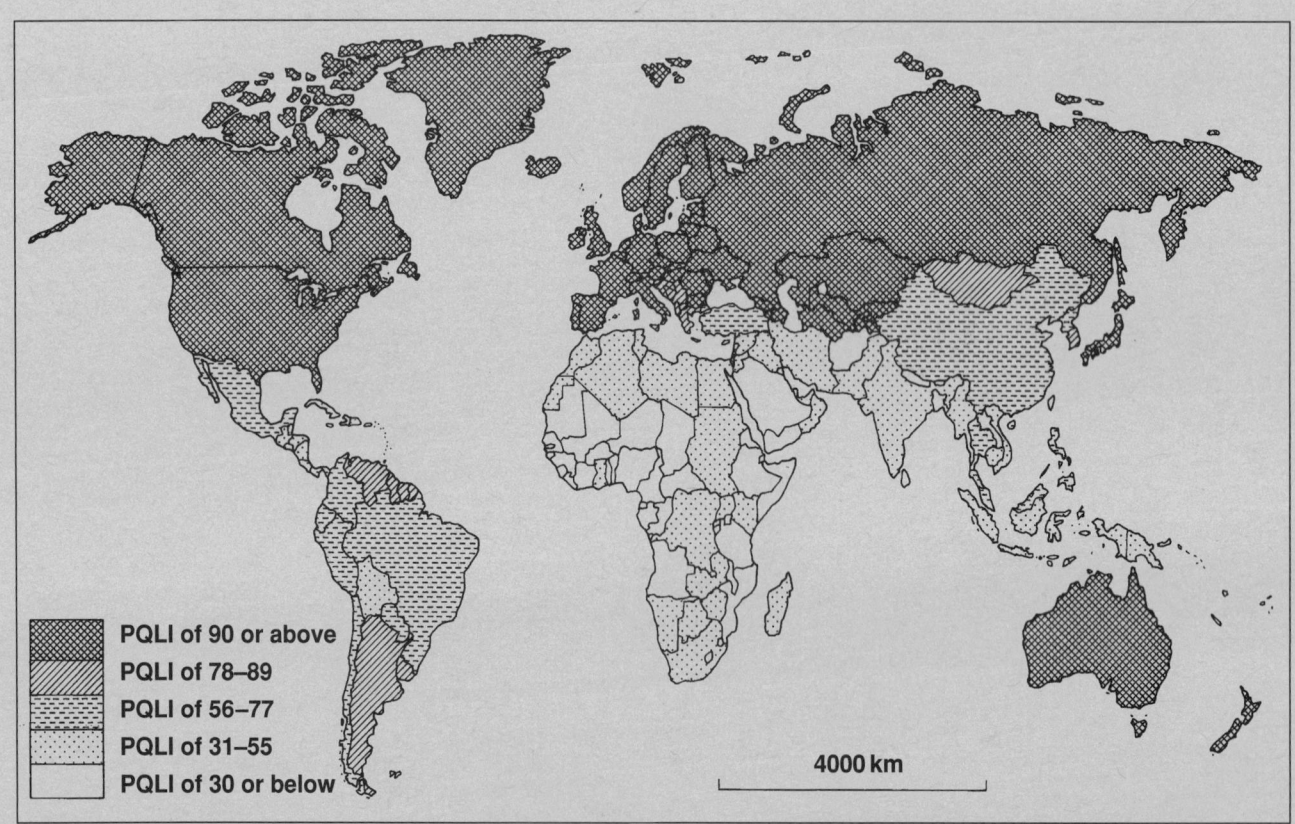

- PQLI of 90 or above
- PQLI of 78–89
- PQLI of 56–77
- PQLI of 31–55
- PQLI of 30 or below

4000 km

[END OF SPECIMEN QUESTION PAPER]

2005 | Higher

[BLANK PAGE]

X208/301

NATIONAL
QUALIFICATIONS
2005

MONDAY, 16 MAY
9.00 AM – 10.30 AM

GEOGRAPHY
HIGHER
Paper 1
Physical and
Human Environments

Six questions should be attempted, namely:

all four questions in **Section A** (Questions 1, 2, 3 and 4);

one question from **Section B** (Question 5 **or** Question 6);

one question from **Section C** (Question 7 **or** Question 8).

Write the numbers of the **six** questions you have attempted in the marks grid on the back cover of your answer booklet.

The value attached to each question is shown in the margin.

Credit will be given for appropriate maps and diagrams, and for reference to named examples.

Marks may be deducted for bad spelling, bad punctuation and for writing that is difficult to read.

Note   The reference maps and diagrams in this paper have been printed in black only: no other colours have been used.

SCOTTISH
QUALIFICATIONS
AUTHORITY

©

1:50 000 Scale
Landranger Series

Four colours should appe
Four colours should appe

Diagrammatic only

2 centimetr

1 kilometre = 0·6214 mile

Extract No 1409/89

ase return to the invigilator.
ase return to the invigilator.

1 mile = 1·6093 kilometres

Marks

SECTION A: Answer ALL questions in this section

Question 1: Lithosphere

Study OS Map Extract number 1409/89: Wast Water (*separate item*), and Reference Map Q1.

(*a*) **Describe** the evidence which shows that Area A, shown on Reference Map Q1, has been affected by the processes of glacial erosion. (You should refer to specific named features and make use of grid references.)   3

(*b*) Choose any **two** features of glacial erosion described in your answer to part (*a*) and, with the aid of annotated diagrams, **explain** how they were formed.   6

Reference Map Q1

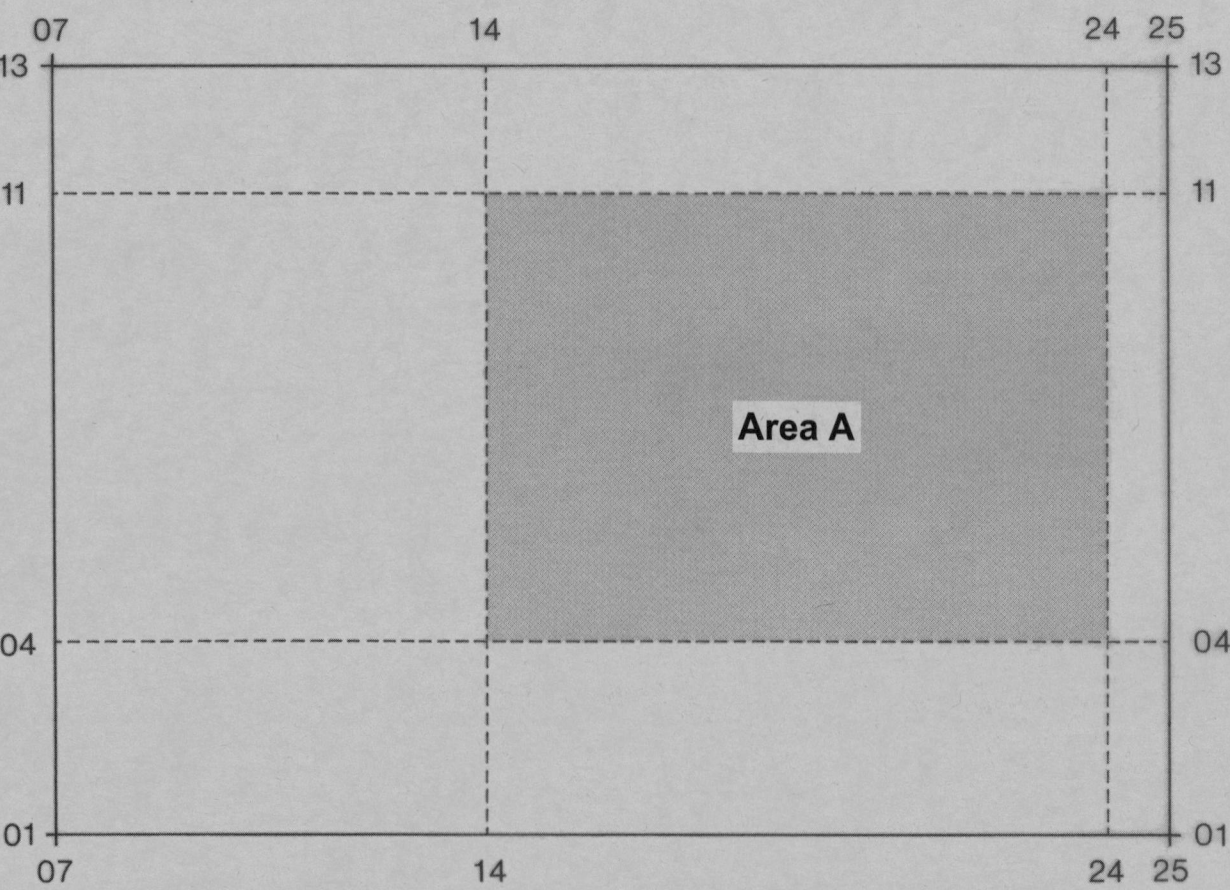

Marks

Question 2: Biosphere

(*a*) **Explain** fully what is meant by the term climax vegetation.   3

(*b*) Study Reference Map Q2 and Reference Diagram Q2 which show a coastal sand dune area.

Describe and **account for** the different plant types likely to be found at locations A, B and C, as you move inland from the coast.   6

Reference Map Q2 (A coastal sand dune area)

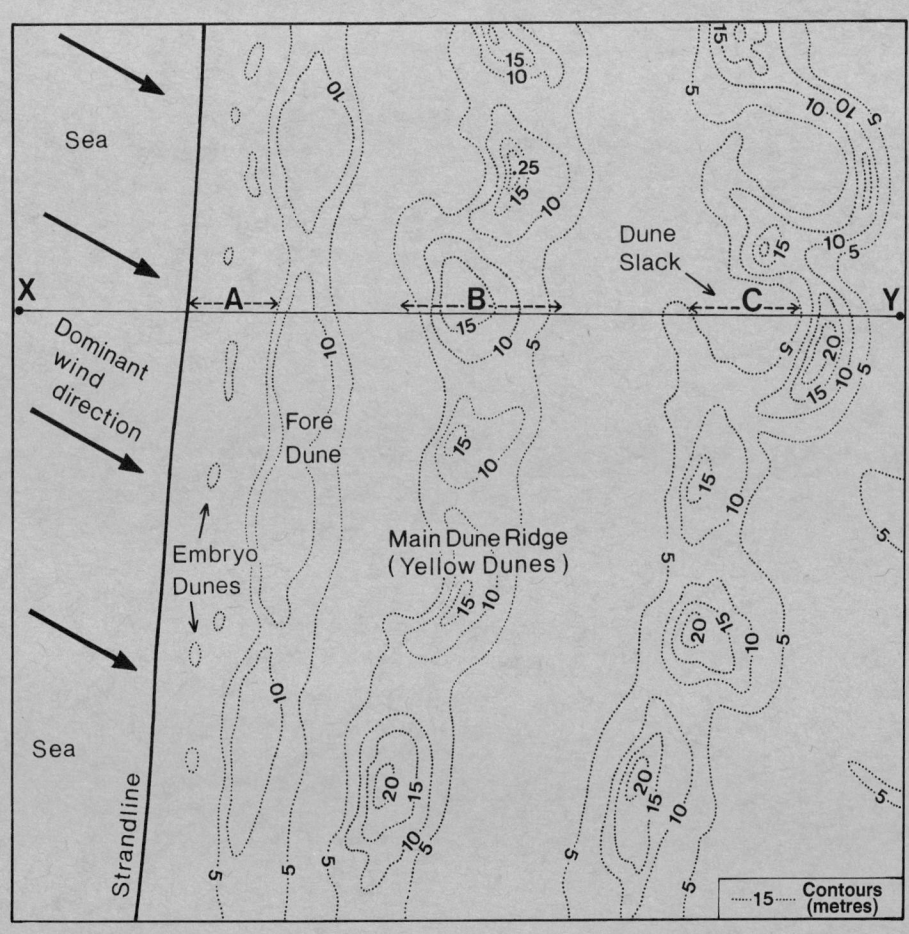

Reference Diagram Q2 (Cross-section from X to Y on the map above)

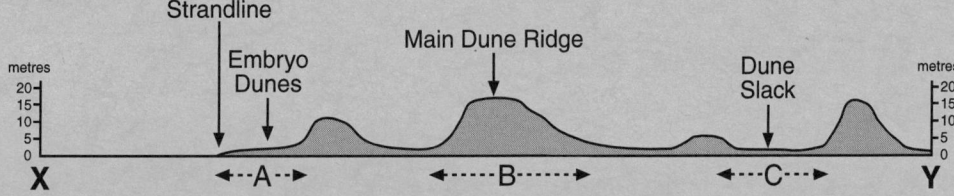

[Turn over

Marks

Question 3: Population Geography

Study Reference Diagram Q3.

(a) **Describe** and **explain** the population structure of Botswana in **2000**.

5

(b) **Describe** the population structure of **2025** and **suggest reasons** why it is expected to be so different. Note that one age group is tracked on both diagrams.

4

Reference Diagram Q3 (Age pyramids for Botswana in Southern Africa, 2000 and 2025)

Botswana: 2000

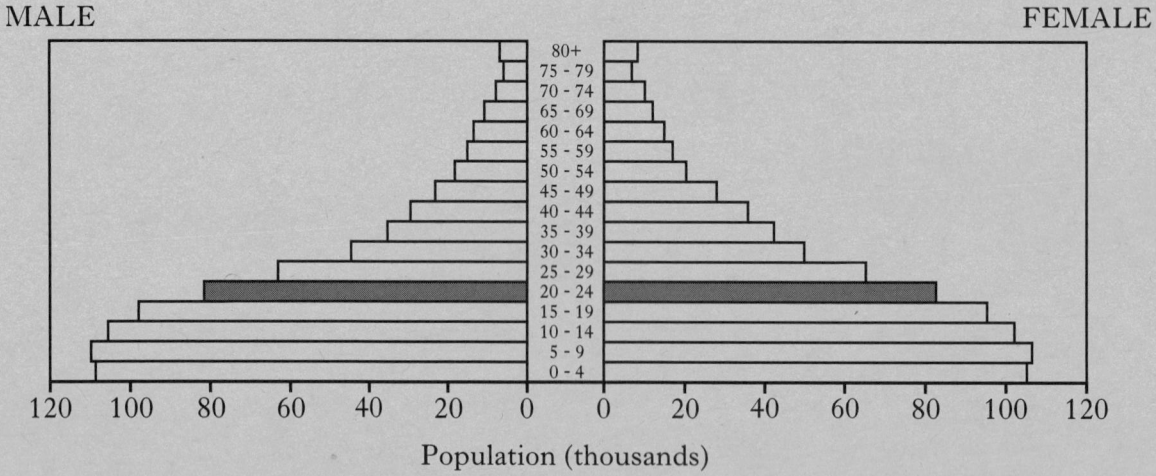

Botswana: 2025

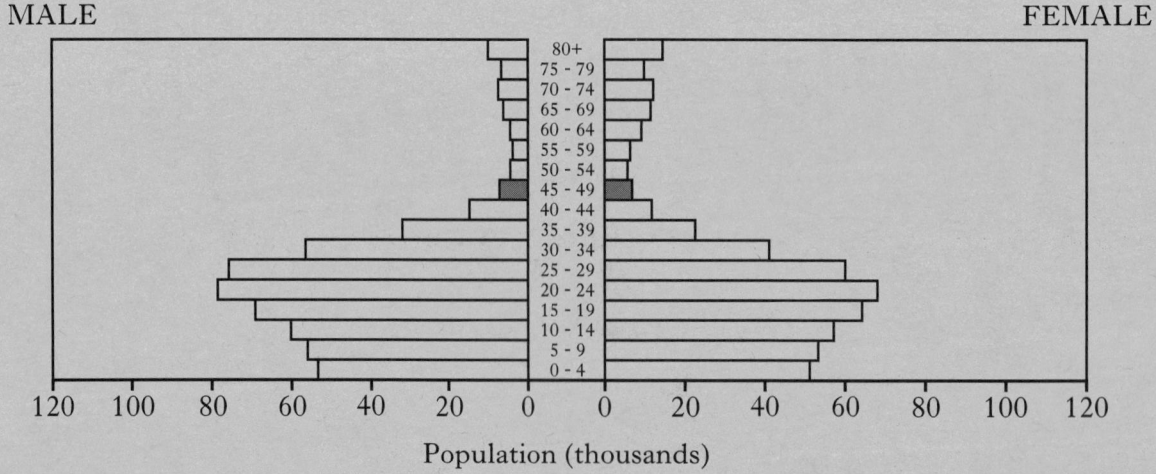

Marks

Question 4: Urban Geography

(*a*) For a city you have studied, **explain** the ways in which its site and situation contributed to its growth.

4

(*b*) Study Reference Map Q4 which shows the redevelopment of an inner city landscape along Glasgow's waterfront.

Referring to an inner city landscape in Glasgow, **or** in a city you have studied, **describe** and **explain** the changes which have taken place in recent years.

5

Reference Map Q4 (Planned redevelopment of part of Glasgow)

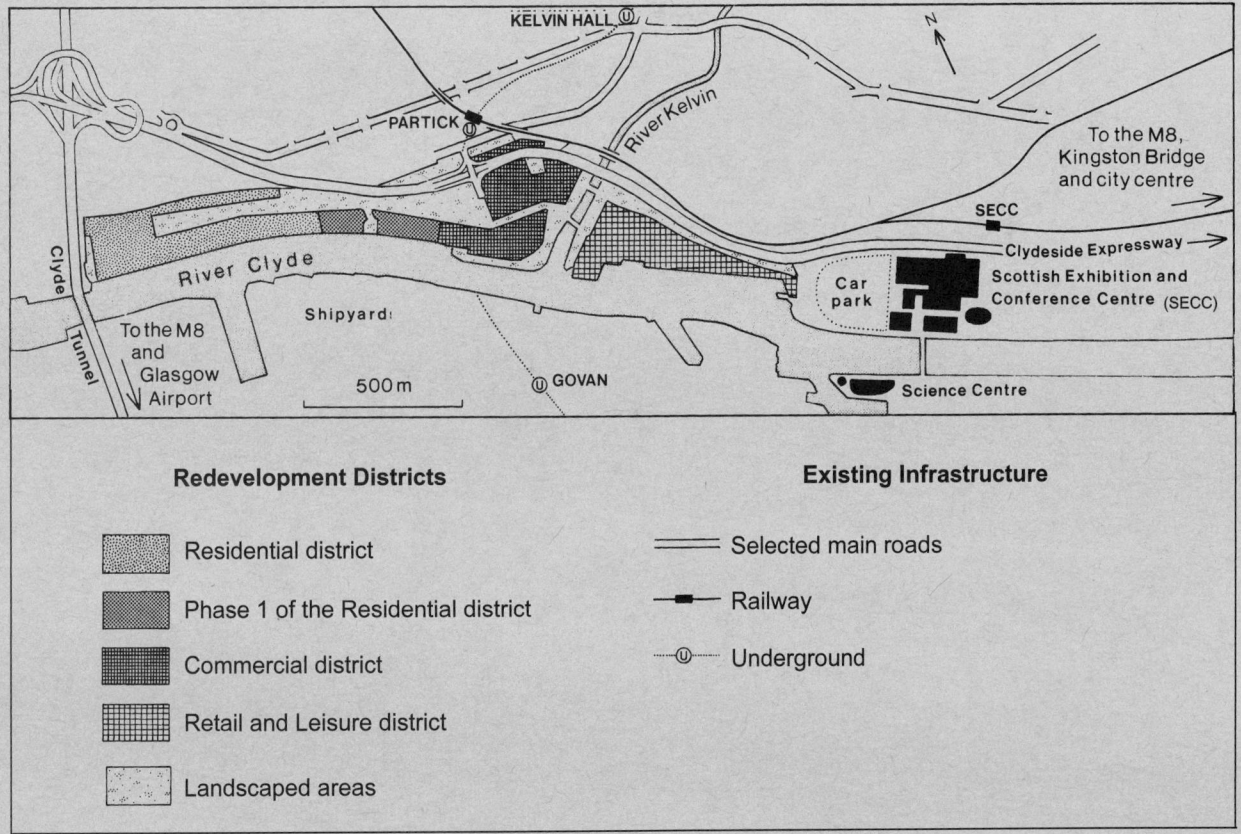

[Turn over

Marks

**SECTION B: Answer ONE question from this section,
ie either Question 5 or Question 6.**

Question 5: Atmosphere

(*a*) Study Reference Diagram Q5.

Explain the **human** factors which may have led to the changes in global mean temperature shown in the diagram.

4

(*b*) With the aid of an annotated diagram, **explain** why Tropical latitudes receive more of the sun's energy than Polar regions.

3

Reference Diagram Q5 (Changes in global mean temperature 1880–2000)

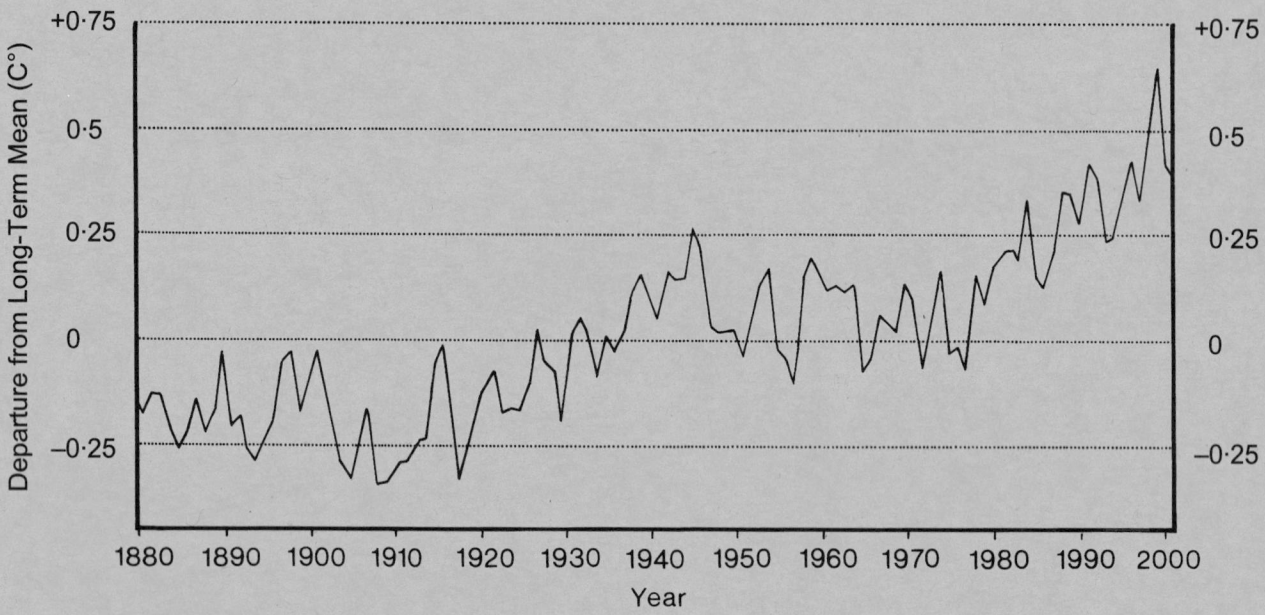

Marks

**DO NOT ANSWER THIS QUESTION IF YOU HAVE
ALREADY ANSWERED QUESTION 5**

Question 6: Hydrosphere

Study OS Map Extract number 1409/89: Wast Water (*separate item*).

(*a*) Using appropriate grid references, **describe** the physical characteristics of the River Bleng and its valley from 136095 (Stockdale Head) to 102032 (where it meets the River Irt).

4

(*b*) The River Irt has a number of meanders on the stretch of river from 145039 (Wast Water) to 104010.

Explain, with the aid of a diagram or diagrams, how a meander is formed.

3

[Turn over

Marks

SECTION C: Answer ONE question from this section, ie either Question 7 or Question 8.

Question 7: Rural Geography

Study Reference Diagram Q7 which illustrates the "traditional" features of an intensively farmed area in South-east Asia.

(*a*) **Describe** and **account for** the main features of the farming landscape shown on the sketch.

4

(*b*) **Outline** the changes in farming practices which may have taken place in recent years.

3

Reference Diagram Q7 (An intensive peasant farming landscape in SE Asia)

Marks

DO NOT ANSWER THIS QUESTION IF YOU HAVE ALREADY ANSWERED QUESTION 7

Question 8: Industrial Geography

(a) | *"Traditional industries were often located on or near raw material sources."* |

For any industrial concentration in the EU which you have studied, **describe** the **physical** factors which led to the growth of industry before 1950.

3

(b) Study Reference Map Q8.

Describe and **explain** the **human** factors which have led modern industries to locate along the corridor between London and Peterborough.

4

Reference Map Q8 (A growing industrial area of the UK)

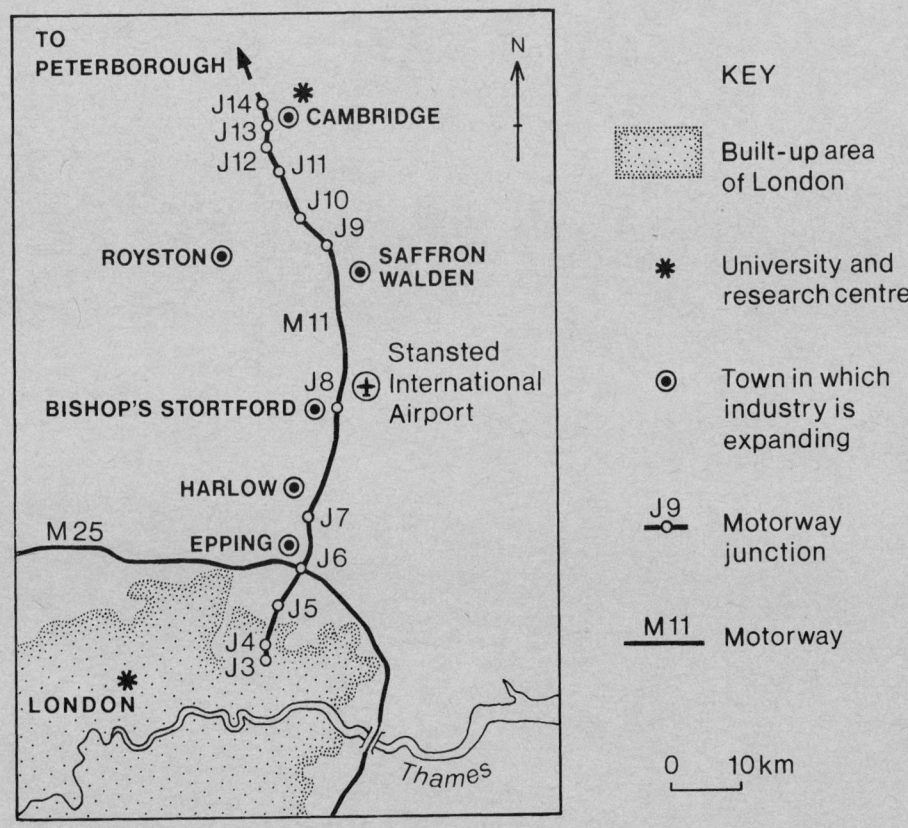

[END OF QUESTION PAPER]

[BLANK PAGE]

X208/303

NATIONAL
QUALIFICATIONS
2005

MONDAY, 16 MAY
10.50 AM – 12.05 PM

GEOGRAPHY

HIGHER
Paper 2
Environmental
Interactions

Two questions should be attempted, namely:

one question from **Section 1** (Questions 1, 2, 3) and
one question from **Section 2** (Questions 4, 5, 6).

Write the numbers of the **two** questions you have attempted in the marks grid on the back cover of your answer booklet.

The value attached to each question is shown in the margin.

Credit will be given for appropriate maps and diagrams, and for reference to named examples.

Marks may be deducted for bad spelling, bad punctuation and for writing that is difficult to read.

Questions should be answered in sentences.

Note   The reference maps and diagrams in this paper have been printed in black only: no other colours have been used.

SCOTTISH
QUALIFICATIONS
AUTHORITY

THB   X208/303   6/15370

Marks

SECTION 1

You must answer ONE question from this Section.

Question 1 (Rural Land Resources)

(a) Study Reference Map Q1 and Reference Table Q1.

   Explain why the number of visitors to National Parks can vary so greatly.   **5**

(b) For any National Park **or** protected Upland area you have studied,

   (i)   **explain** how the area may benefit from an influx of tourists, and

   (ii) **EITHER**

   describe how the increased volume of traffic is managed.

   OR

   describe how fragile environments may be conserved.   **9**

(c) | *"The Peak District is one of the National Parks famous for its Carboniferous Limestone Scenery."* |
     | --- |

With the aid of annotated diagrams, **describe** and **explain** how the main features of Carboniferous Limestone were formed. Both surface and underground features should be included in your answer.   **11**

(25)

Question 1 – continued

Reference Map Q1 (National Parks in England and Wales)

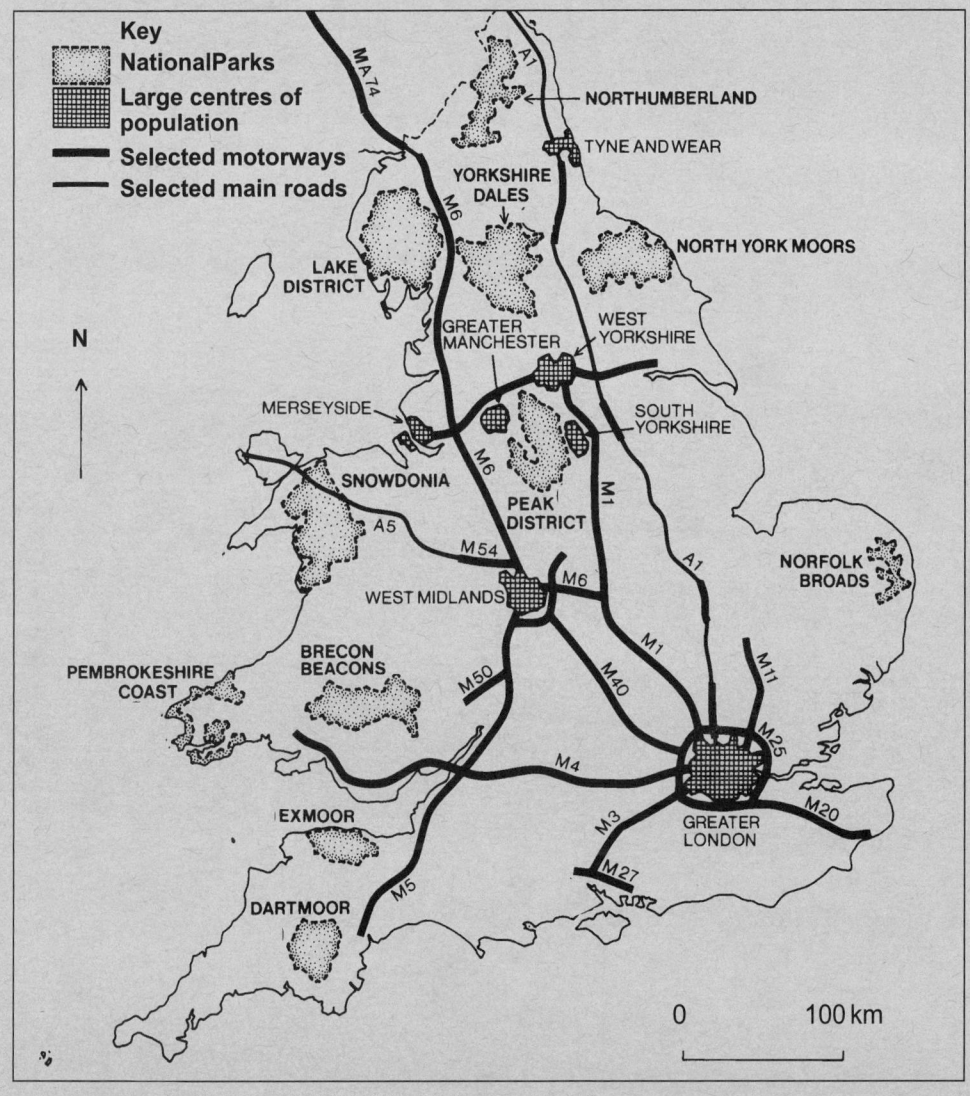

Reference Table Q1 (General National Park Statistics)

	Brecon Beacons	Dartmoor	Exmoor	Lake District	Northumberland	North York Moors	Peak District	Pembrokeshire Coast	Snowdonia	Yorkshire Dales	Norfolk Broads
Designation Year	1957	1951	1954	1951	1956	1952	1951	1952	1951	1954	1989
Area (ha)	135 144	95 570	69 280	229 198	104 947	143 603	143 833	62 000	214 159	176 869	30 292
Visitor days (millions per year)	7	8	3	20	1·5	11	22	13	8	9	5

[Turn over

Marks

Question 2 (Rural Land Degradation)

(*a*) **Describe** and **explain** the processes of wind erosion shown in Reference Diagram Q2.

3

(*b*) **Describe** and **explain** the human activities which have led to the degradation of rural land in **either** Africa north of the Equator **or** the Amazon Basin.

8

(*c*) Referring to named locations in **either** Africa north of the Equator **or** the Amazon Basin, **describe** the impact of land degradation on the landscape and people.

6

(*d*)

> "*Areas of **North America** have suffered from rural land degradation.*"

    (i)   Referring to named areas in **North America** that you have studied, **describe** and **explain** the ways in which farmers have adjusted their farming methods to reduce the risk of soil erosion.

    (ii) **Comment** on the effectiveness of these methods.

8

(25)

Question 2 – continued

Reference Diagram Q2 (Selected processes of wind erosion)

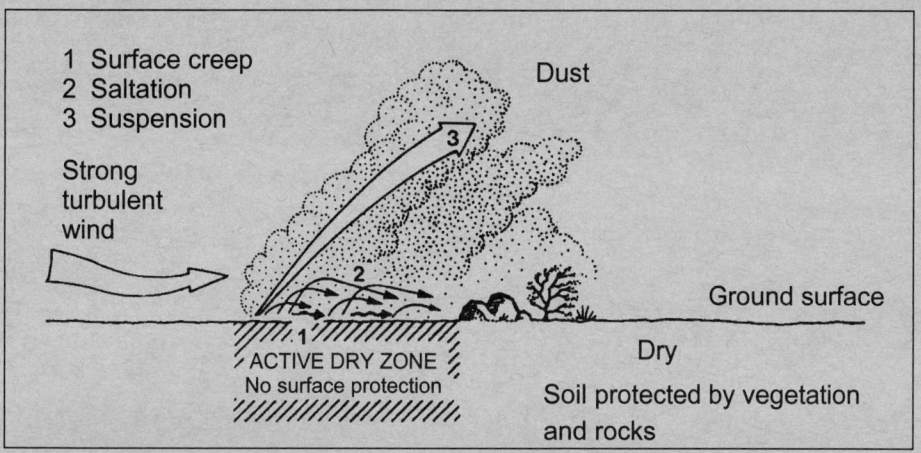

[Turn over

Question 3 (River Basin Management)   *Marks*

Study Reference Maps Q3A, Q3B and Q3C.

(*a*) **Explain** why there is a need for water management in Ghana   **5**

(*b*) The Akosombo Dam lies at the southern end of Lake Volta.

For the Akosombo Dam **or** any dam you have studied in Africa **or** North
America **or** Asia, **explain** the **physical** factors which have to be considered
when selecting the site for the dam.   **5**

(*c*) **Describe** and **account for** the social, economic and environmental
benefits **and** adverse consequences of a named major water control project
in Africa **or** North America **or** Asia.   **12**

(*d*) For the Volta River Project **or** any other water control project, **explain** any
political problems which may have resulted from the development.   **3**

(**25**)

Reference Map Q3A (Map of Ghana and Lake Volta)

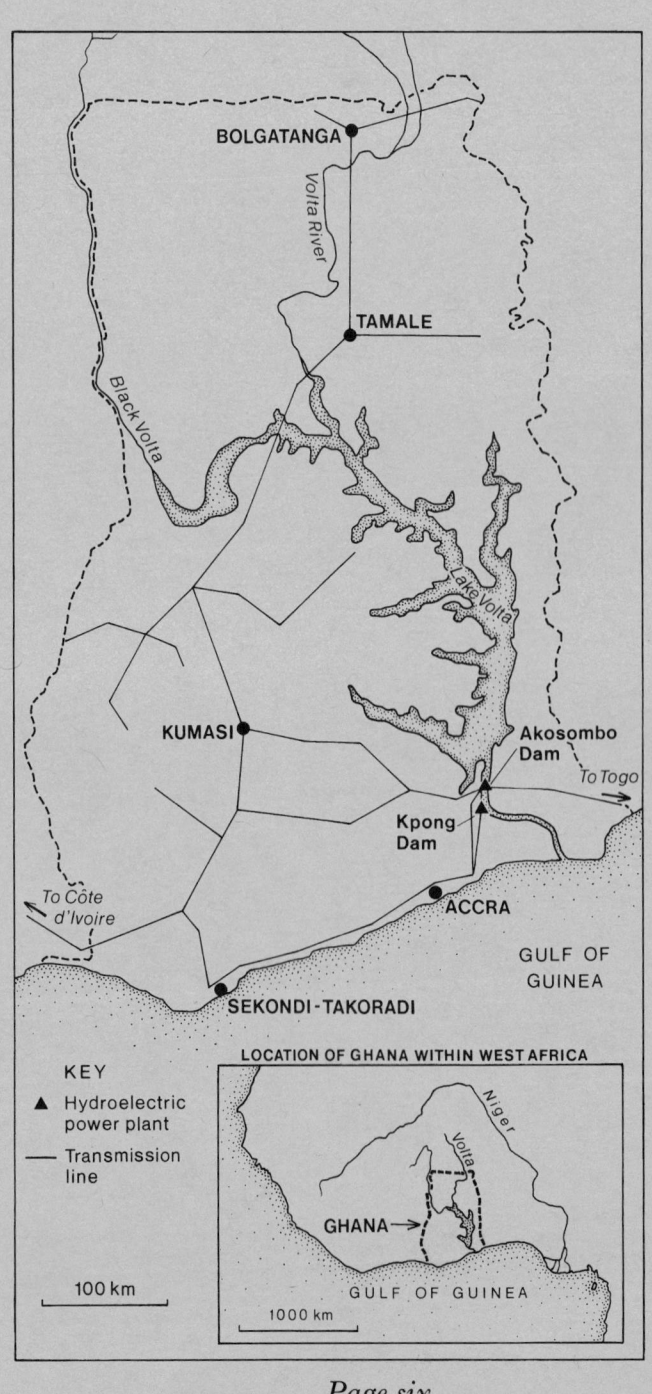

Question 3 – continued

Reference Map Q3B (Mean annual temperatures in Ghana)

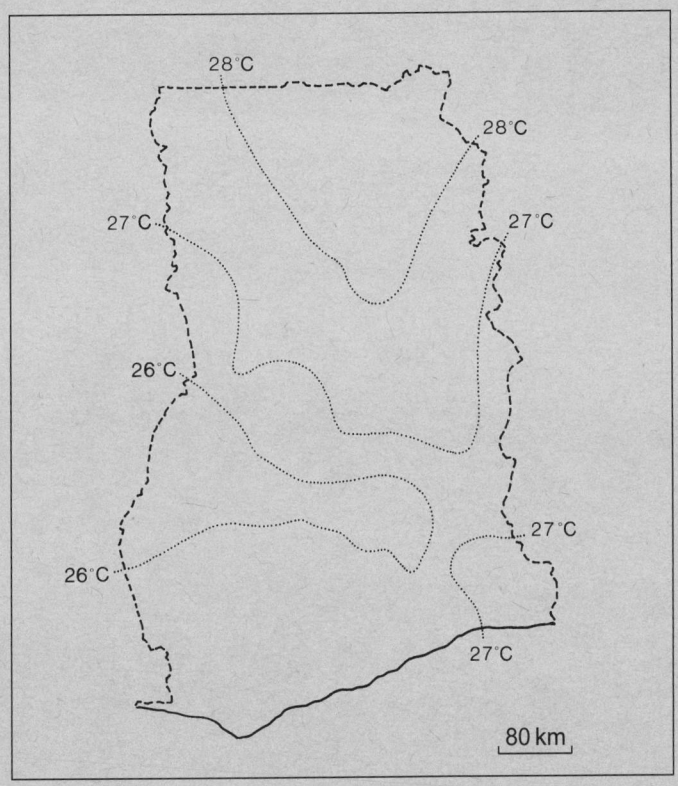

Reference Map Q3C (Annual rainfall in Ghana)

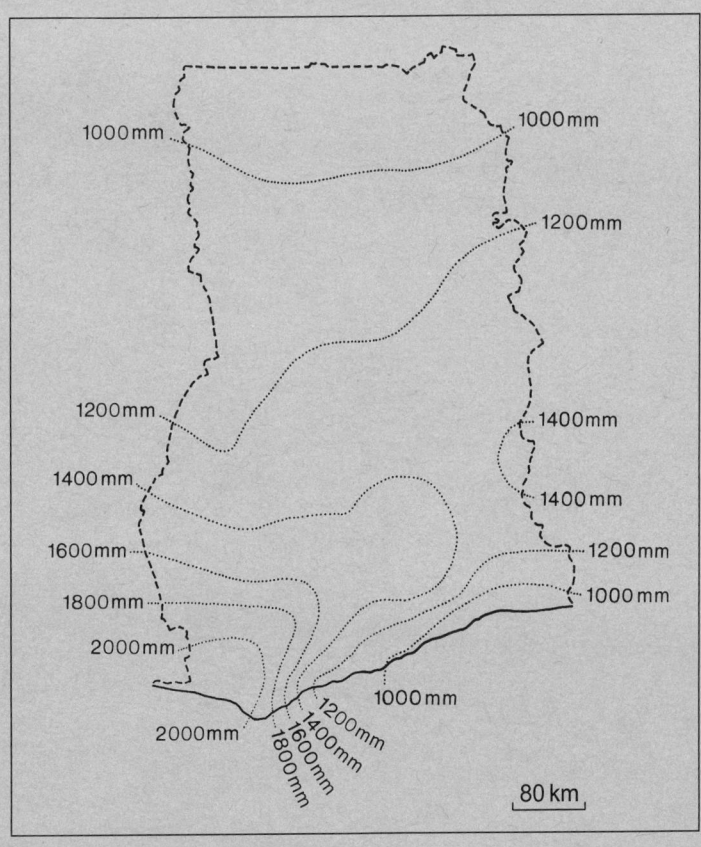

Marks

SECTION 2

You must answer ONE question from this Section.

Question 4 (Urban Change and its Management)

> *"The world is facing explosive growth of urban population, mainly in the Developing world".*

(a) With reference to cities you have studied, **suggest** why the populations of cities in the ELDCs (Economically Less Developed Countries) are forecast to grow much more rapidly than those of cities in the EMDCs (Economically More Developed Countries).

6

(b) With reference to a named city you have studied in an ELDC,

   (i) **describe** the social, economic and environmental problems created by its rapid growth,

   (ii) **describe** some of the methods used to tackle these problems, and

   (iii) **comment** on the effectiveness of the methods used.

10

(c) Reference Map Q4 identifies London's congestion charging zone, a scheme which attempts to solve traffic problems in the capital.

   (i) **Suggest** reasons why the city centre of London, **or** a named city you have studied in an EMDC, suffers from traffic congestion.

   (ii) **Describe** the different ways the city has attempted to solve traffic problems and **discuss** the extent to which these methods have been successful.

9

(25)

Question 4 - continued

Reference Map Q4 (London's congestion charging zone)

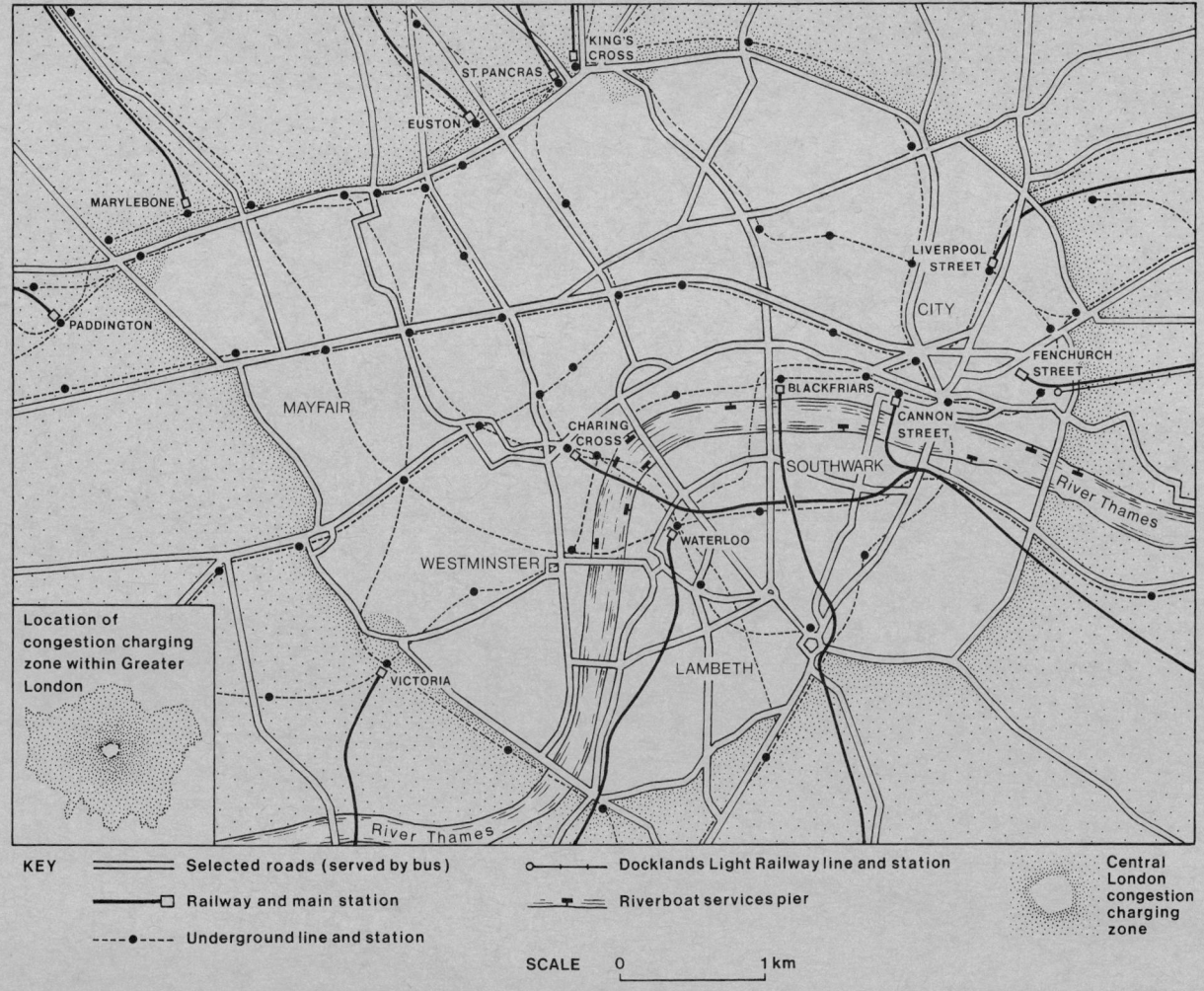

KEY

⟨▤⟩ Selected roads (served by bus)

⟨▭⟩ Railway and main station

⟨- - -•- - -⟩ Underground line and station

Location of congestion charging zone within Greater London

⟨∘─•─⟩ Docklands Light Railway line and station

⟨▭▬▭⟩ Riverboat services pier

SCALE 0   1 km

Central London congestion charging zone

[Turn over

Marks

Question 5 (European Regional Inequalities)

Study Reference Map Q5A.

(a) **Describe** the distribution of those regions which are eligible for European Union funds under Objective 1 support (2000–2006).

4

(b) Some regions have benefited in the past from Objective 1 support. They are currently being "phased out" of the programme and are being given "transitional support". **Describe** the benefits these regions received under Objective 1 support.

5

Study Reference Map Q5B and Reference Table Q5.

(c) To what extent does the data provide evidence of regional inequalities within Germany?

6

(d) For **either** Germany **or** a named country you have studied in the European Union,

  (i) **describe** and **explain** both the physical **and** human factors that have led to regional inequalities, and

7

  (ii) **describe** the steps taken by the national government to tackle problems in less prosperous regions.

3

(25)

Reference Map Q5A
(European Union Objective 1 funding)

Reference Map Q5B
(Regions in Germany)

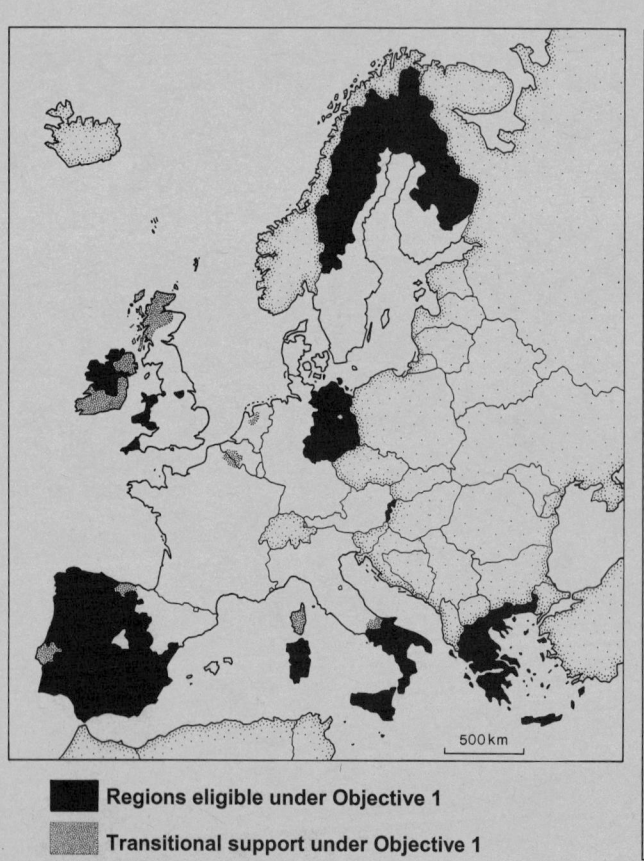

◼ Regions eligible under Objective 1

▦ Transitional support under Objective 1

▧ Non EU countries (2003)

Question 5 – continued

Reference Table Q5 (Selected socio-economic statistics: Germany)

Regions	Population (000)	GDP (Billion Euro)	Gross monthly earnings (2002 average/Euro)	% Unemployment	Hospital Beds per 000
Baden-Württemberg	10 601	307·44	3369	4·4	9·13
Bavaria	12 330	368·92	3363	4·6	9·73
Berlin	3388	77·13	3141	15·6	6·96
Brandenburg	2593	44·12	2415	16·9	8·45
Bremen	660	22·96	3420	10·1	9·65
Hamburg	1726	75·18	3492	8·2	7·61
Hesse	6078	191·61	3487	5·9	9·90
Mecklenburg-Western Pomerania	1760	29·61	2194	19·1	12·47
Lower Saxony	7956	183·12	3048	7·2	8·48
North Rhine-Westphalia	18 052	463·96	3216	7·2	8·71
Rhineland-Palatinate	4049	93·30	3140	5·6	8·81
Saarland	1066	25·43	2977	7·6	10·71
Saxony	4384	75·79	2408	17·8	8·80
Saxony-Anhalt	2581	43·31	2395	19·2	8·47
Schleswig-Holstein	2804	65·64	2964	7·6	10·18
Thuringia	2411	40·67	2333	15·1	10·12
Germany	**82 440**	**2108·20**	**3198**	**10·75**	**9·09**

[Turn over for Question 6 on *Page twelve*

Marks

Question 6 (Development and Health)

(a) Study Reference Map Q6 which shows the Human Development Index (HDI) for countries of the world.

   (i) For **either** the HDI **or** any similar composite measure of development, state **three** indicators which may be used in its calculation, and **comment** on the usefulness of each of these indicators.

6

   (ii) Referring to **one** ELDC (Economically Less Developed Country) which you have studied, **suggest reasons** for the variations in levels of development which exist **within** that country.

6

(b) For **either** malaria **or** bilharzia (schistosomiasis) **or** cholera,

   (i) **describe** and **evaluate** the strategies used in controlling the spread of the disease, and

   (ii) **explain** the benefits of controlling the disease to those countries affected by it.

13

(25)

Reference Map Q6 (The World: Human Development Index (HDI))

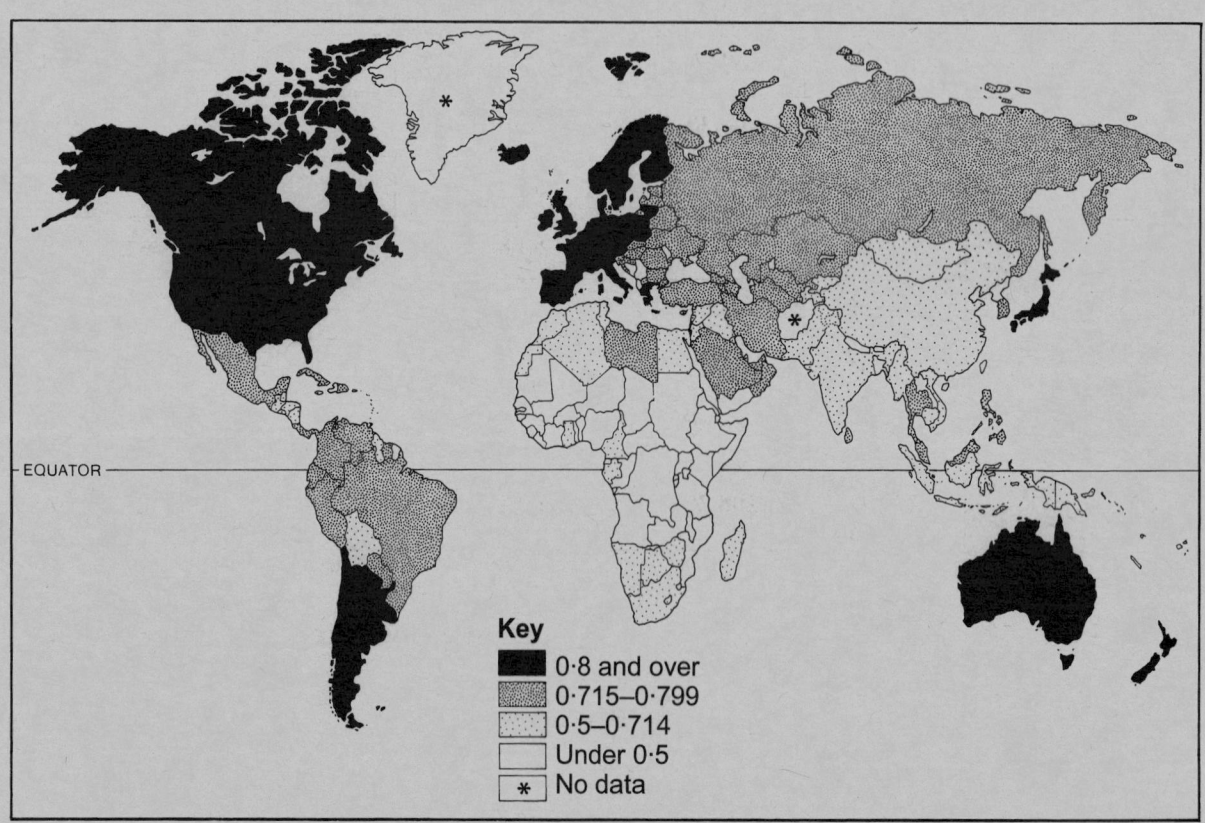

Key
- 0·8 and over
- 0·715–0·799
- 0·5–0·714
- Under 0·5
- * No data

[END OF QUESTION PAPER]

[BLANK PAGE]

Acknowledgements

Leckie & Leckie is grateful to the copyright holders, as credited, for permission to use their material:
Nelson Thornes Ltd for an extract from *Themes & Issues – 1997 – National Parks in the UK* (2002 Applications p 2);
Harcourt Education for a graph from *AS Level Geography* by A. Bowen & J. Pallister (2002 Applications p 10).
This product includes mapping data reproduced by permission of Ordnance Survey on behalf of HMSO © Crown copyright 2005. License no. 100036009.

The following companies/individuals have very generously given permission to reproduce their copyright material free of charge:
Causeway Press Ltd for an extract from *Geography in Focus* by Ian Cook (2002 Applications p 5);
The Population Reference Bureau for a table (2002 Applications p 10);
BC Hydro for an extract (2003 Applications p 5);
POPIN for a chart from *UN World Urbanization Prospects: The 1999 Revision* (2003 Applications p 6);
The Geographical Association for an extract from *Discovering Cities* by CM Law (2003 Applications p 7);
Hodder & Stoughton for an extract from *Core Higher Geography* by McLean and Thomson (2003 Applications p 8).